ENERGY 2021

The Eleventh International Conference on Smart Grids, Green Communications and IT Energy-aware Technologies

May 30th – June 3rd, 2021

ENERGY 2021 Editors

Vivian Sultan, California State University Los Angeles, USA

Eric MSP Veith, OFFIS e.V. - Oldenburg, Germany

Printed from e-media with permission by:

Curran Associates, Inc.
57 Morehouse Lane
Red Hook, NY 12571

Some format issues inherent in the e-media version may also appear in this print version.

Printed with permission by Curran Associates, Inc. (2023)

International Academy, Research, and Industry Association (IARIA)
412 Derby Way
Wilmington, DE 19810

Phone: (408) 893-6407
Fax: (408) 527-6351

petre@iaria.org

Additional copies of this publication are available from:

Curran Associates, Inc.
57 Morehouse Lane
Red Hook, NY 12571 USA
Phone: 845-758-0400
Fax: 845-758-2633
Email: curran@proceedings.com
Web: www.proceedings.com

Table of Contents

ENERGY 2021

Foreword

The Eleventh International Conference on Smart Grids, Green Communications and IT Energy-aware Technologies (ENERGY 2021), held between May 30 – June 3rd, 2021, continued the event considering Green approaches for Smart Grids and IT-aware technologies. It addressed fundamentals, technologies, hardware and software needed support, and applications and challenges.

There is a perceived need for a fundamental transformation in IP communications, energy-aware technologies and the way all energy sources are integrated. This is accelerated by the complexity of smart devices, the need for special interfaces for an easy and remote access, and the new achievements in energy production. Smart Grid technologies promote ways to enhance efficiency and reliability of the electric grid, while addressing increasing demand and incorporating more renewable and distributed electricity generation. The adoption of data centers, penetration of new energy resources, large dissemination of smart sensing and control devices, including smart home, and new vehicular energy approaches demand a new position for distributed communications, energy storage, and integration of various sources of energy.

We take here the opportunity to warmly thank all the members of the ENERGY 2021 Technical Program Committee, as well as the numerous reviewers. The creation of such a high quality conference program would not have been possible without their involvement. We also kindly thank all the authors who dedicated much of their time and efforts to contribute to ENERGY 2021. We truly believe that, thanks to all these efforts, the final conference program consisted of top quality contributions.

Also, this event could not have been a reality without the support of many individuals, organizations, and sponsors. We are grateful to the members of the ENERGY 2021 organizing committee for their help in handling the logistics and for their work to make this professional meeting a success.

We hope that ENERGY 2021 was a successful international forum for the exchange of ideas and results between academia and industry and for the promotion of progress in the fields of smart grids, green communications and IT energy-aware technologies.

We are convinced that the participants found the event useful and communications very open.

ENERGY 2021 Chairs:

Vincenzo Gulisano, Chalmers University of Technology, Sweden
Marco Lützenberger, Technische Universität Berlin, Germany
Evangelos Pournaras, ETH Zurich, Switzerland
Chun-Hsi Huang, University of Connecticut, USA
Magnus Almgren, Chalmers University of Technology, Sweden

ENERGY 2021 Publicity Chairs

Daniel Basterretxea, Universitat Politecnica de Valencia, Spain
Marta Botella-Campos, Universitat Politecnica de Valencia, Spain

ENERGY 2021

Committee

ENERGY 2021 Steering Committee

Eric MSP Veith, OFFIS e.V. – Oldenburg, Germany
Dragan Obradovic, Siemens - Corporate Technology, Munich, Germany
Mark Apperley, University of Waikato, New Zealand
Michael Negnevitsky, University of Tasmania, Australia
Vivian Sultan, California State University Los Angeles, USA
Steffen Fries, Siemens, Germany

ENERGY 2021 Industry/Research Advisory Committee

Hongbo Sun, Mitsubishi Electric Research Laboratories, USA
Daisuke Mashima, Advanced Digital Sciences Center, Singapore
Sambaran Bandyopadhyay, IBM Research, India
Chresten Træholt, Center for Electric Power and Energy - CEE | Technical University of Denmark - DTU, Denmark
Vincenzo Gulisano, Chalmers University of Technology, Sweden
Marco Lützenberger, Technische Universität Berlin, Germany
Evangelos Pournaras, ETH Zurich, Switzerland
Chun-Hsi Huang, University of Connecticut, USA
Magnus Almgren, Chalmers University of Technology, Sweden

ENERGY 2021 Publicity Chairs

Daniel Basterretxea, Universitat Politecnica de Valencia, Spain
Marta Botella-Campos, Universitat Politecnica de Valencia, Spain

ENERGY 2021 Technical Program Committee

Santosh Aditham, University of South Florida, USA
Kodjo Agbossou, Université du Québec à Trois-Rivières, Canada
Mohammad Pourmahmood Aghababa, Urmia University of Technology, Iran
Kamal Al-Haddad, École de technologie supérieure, Montreal, Canada
Ahmed Al-Salaymeh, The University of Jordan, Amman, Jordan
Miltos Alamaniotis, University of Texas at San Antonio, USA
Magnus Almgren, Chalmers University of Technology, Sweden
Mark Apperley, University of Waikato, New Zealand
Ashrant Aryal, Texas A&M University, USA
Navid Bayati, Aalborg University, Denmark
Andreas Berl, Technische Hochschule Deggendorf, Germany
Vito Calderaro, University of Salerno, Italy
M. Girish Chandra, TCS Research & Innovation, India

Copyright Information

Control of Synchronization in Two-Layer Power Grids

Simona Olmi
Istituto dei Sistemi Complessi
CNR - Consiglio Nazionale delle Ricerche
Sesto Fiorentino, Italy
email:simona.olmi@fi.isc.cnr.it
ORCID 0000-0002-8272-3493

Carl H. Totz
Institut für Theoretische Physik
Technische Universität Berlin
Berlin, Germany
email:carl.hendrik@web.de

Eckehard Schöll
Institut für Theoretische Physik
Technische Universität Berlin
Berlin, Germany
Potsdam Institute for Climate Impact Research
Potsdam, Germany
email:schoell@physik.tu-berlin.de
ORCID 0000-0002-7318-2672

Abstract—**In this work, we suggest to model the dynamics of power grids in terms of a two-layer network, and use the Italian high voltage power grid as a proof-of-principle example. The first layer in our model represents the power grid consisting of generators and consumers, while the second layer represents a dynamic communication network that serves as a controller of the first layer. In particular, the dynamics of the power grid is modelled by the Kuramoto model with inertia, while the communication layer provides a control signal P_i^c for each generator to improve frequency synchronization within the power grid. We propose different realizations of the communication layer topology and different ways to calculate the control signal. Then, we conduct a systematic survey of the two-layer system against a multitude of different realistic perturbation scenarios, such as disconnecting generators, increasing demand of consumers, or generators with stochastic power output. When using a control topology that allows all generators to exchange information, we find that a control scheme aimed to minimize the frequency difference between adjacent nodes operates very efficiently even against the worst scenarios with the strongest perturbations.**
Keywords-nonlinear complex networks; power grids; synchronization; stability analysis; control

I. Introduction

Global warming, the growing world population and power demand, with a subsequent increase in carbon power emissions, have provoked governments and energy utilities to take solid steps towards the use of renewable energies [1] and their integration within the existing power transmission and distribution systems, thus challenging scientific and technological research towards the goal of increasing the efficiency and flexibility of the power system [2]–[5]. The existing power grid was developed using a centralistic approach, therefore we have a few very high-power ac plants operating at 50 or 60 Hz interconnected by ac or dc transmission systems operating at very high voltages (e.g., 400 kV) and many substations, where the high voltage is transformed to the distribution level (e.g., 20 kV). In order to distribute the power in a capillary way, a huge number of distribution lines is present, supplying the loads directly (in the case of high-power loads) or after voltage transformation in the case of residential or low-power industrial loads (e.g., 400 V in Europe). Recently, renewable energy generators, which produce a few kilowatts in the case of residential photovoltaic systems, up to some megawatts in the case of large photovoltaic and wind generators, have become widely dispersed around the world, thus transforming the present power system into a large-scale distributed generation system incorporating thousands of generators, characterized by different technologies, voltage, current, and power levels, as well as topologies [6] [7]. Hence, their integration with the existing network is fundamentally changing the whole electrical power system [8] [3]: the drawback of renewable energy power plants is that their output is subject to environmental fluctuations outside of human control, i.e., clouds blocking the sun or lack of wind, and these fluctuations emerge on all timescales displaying non-Gaussian behaviour [9] [10]. In addition, these issues are further complicated by the aging infrastructure of the existing power grid, which already cause problems to utilities and customers, providing low power quality at increasing cost. In particular, the power grid infrastructure is very critical and contains a large number of interconnected components: generators, power transformers, and distribution feeders that are geographically spread. Moreover, its increasing complexity and geographical spread, and the side effects caused by the high penetration of renewable, stochastically fluctuating energy generators make it very vulnerable, both from the point of view of required sophisticated security mechanisms [11] and from the point of view of dynamic stability, since renewable sources are usually employed by microgrids in isolated modes to maintain their capability of connecting and disconnecting from the grid [12]. Due to the design of the current power grid as a centralized system where the electric power flows unidirectional through transmission and distribution lines from power plants to the customer, the control is concentrated in central locations and only partially in substations, while remote ends, like loads, are almost or totally passive. Therefore, it is necessary to design new systems that provide more effective and widely distributed intelligent control embedded in local electricity production, two-way electricity and information flows, thus achieving flexible, efficient, economic, and secure power delivery [13]. The new approach, widely known as Smart Grid [14], requires both a complex two-way communication infrastructure, sustaining power flow between intelligent components, and sophisticated computing and information technologies, as well as business applications. The new approach will include grid energy storage, needed for load balancing and for overcoming energy fluctuations caused by the intrinsic nature of renewable energy sources, in addition to preventing widespread power grid cascading failures [15]

[16]. In particular, control is needed in power networks in order to assure stability and to avoid power breakdowns or cascading failures: one of the most important control goals is the preservation of synchronization within the whole power grid. Control mechanisms able to preserve synchronization are ordered by their time scale on which they act: the first second of the disturbance is mainly uncontrolled, and in this case a power plant will unexpectedly shut down with a subsequent shortage of power in the system, energy is drawn from the spinning reserve of the generators. Within the next seconds, the primary control sets on to stabilize the frequency and to prevent a large drop. Finally, to restore the frequency back to its nominal value of 50 (or 60) Hertz, secondary control is necessary. In many recent studies on power system dynamics and stability, the effects of control are completely neglected or only primary control is considered [17]–[22]. This control becomes less feasible if the percentage of renewable power plants increases, due to their reduced inertia [23] [24]. Few studies are devoted to secondary control [25]–[27] and to time-delayed feedback control [28]–[30].

The present extended abstract is organized as follows: In Section II we present the main results, while in Section III a discussion addressing the impact of the proposed research is presented.

II. Proposed Solution

We consider a two-layer network in a full dynamic description. It consists of a power grid layer and a communication layer, which provides the control for the power grid. Each layer is governed by its own dynamics, which is dependent upon the state of the other layer. In particular, the physical topology that relates the interconnection of distributed generators and loads is described by coupled Kuramoto phase oscillators with inertia, closely related to the swing equations [31], while the communication topology, which describes the information flow of the power system control measurements, depends on the information of the neighbors of each node [32]. Starting from the ideal synchronized state, we investigate the effect of multiple different perturbations to which the system is subject, modelling real threats to synchronization of the network, e.g., failure of nodes, increased consumer demand, power plants with stochastically fluctuating output. To describe the fluctuating power output of renewable energy power plants both Gaussian white noise and more realistic intermittent noise have been used (see [33] for more details). For each perturbation, different setups of the communication layer are tested to find an effective control strategy that successfully preserves frequency synchronization against all applied perturbations. As a proof of concept, the Italian high voltage power grid has been considered. In the communication layer we have assumed a selection of different control schemes (control functions f_i^{diff}, f_i^{dir} and f_i^{comb}) and control topologies (adjacency matrices c_{ij}^{loc} and c_{ij}^{ext}). All control schemes take advantage of the second layer by collecting information from adjacent nodes to calculate the control signal. This can be done either in a local setting (c_{ij}^{loc}) where generators possess the same communication links as in the power grid layer, or in an extended control layer topology (c_{ij}^{ext}) where additional communication links between all generators are present. We have tested (i) a control scheme aimed at synchronizing the frequency of the controlled nodes with their neighbors (difference control f^{diff}), (ii) a control scheme aimed at restoring the original synchronization frequency in the neighborhood of the controlled node (direct control f^{dir}), and (iii) a mixed approach combining both (f^{comb}). The only control scheme being able to effectively counteract all of the perturbations is the difference control scheme f^{diff} in the extended control topology, while the direct control has some advantages in the local control topology only. Moreover, the calculation of different topological measures shows that nodes in the power grid layer which are more affected by perturbations are not characterized, in general, by specific topological features. It turns out that the Italian power grid can be divided in two specific parts: the northern, continental part, with a higher average connectivity, which is more resilient to perturbations, and the southern, peninsular part, characterized by a low average connectivity. The elongated structure of the southern part makes it less robust to perturbations.

III. Conclusion

The aim of this work is to investigate the controllability of power networks subject to different kinds of perturbations and to develop novel control concepts considering the communication infrastructure present in the smart grid. Few works have included the communication layer into the synchronization of power networks. Even though the communication infrastructure plays an important role in control and synchronization, preliminary works [34] [35] assume trivial networks, without disconnected nodes, which, however, is of great importance in stabilizing smart grids, due to the necessity of synchronizing grids with isolated generators, microgrids, or even coupled microgrids that can be connected or disconnected to the main grid at any time. Moreover, the inclusion of a communication infrastructure has added new challenges in control and stability [36], where communication constraints emerge, e.g., time-delays, packet losses, sampling and data rate, among others, but, up to now, attention has focussed on sampling problems in order to assure that synchronization is independent on the sampling period [32]. On the other hand, the same two-layer topology, here implemented, has been already investigated in [37] to understand how localized events can present a severe danger to the stability of the whole power grid, by causing a cascade of failures, but without considering the dynamics of the control nodes. Here the focus of our investigation is on the interdependence of the communication network and the power grid: Random failure of a power plant causes the malfunction of connected elements in the communication layer. Communication nodes isolated due to the failure become inert, causing generators connected to them to shut down as well as eventually leading to a far-reaching blackout. In short, our proposed control techniques preserve synchronization for different perturbations [38], thus demonstrating the powerful perspectives of our control approach which considers synchronization of power systems based on the coupled dynamics of the smart grid architecture and the communication infrastructure.

Acknowledgment

Funded by the Deutsche Forschungsgemeinschaft (DFG, German Research Foundation) - Projektnummer 163436311 - SFB 910.

References

[1] United Nations Framework Convention on Climate Change, "Adoption of the Paris Agreement FCCC/CP/2015/L. 9/Rev. 1", 2015, available at http://unfccc.int/resource/docs/2015/cop21/eng/l09r01.pdf.

[2] A. Vaccaro, G. Velotto, and A. Zobaa, "A decentralized and cooperative architecture for optimal voltage regulation in smart grids", IEEE Trans. Ind. Electron., vol. 58, pp. 4593-4602, 2011.

[3] M. Z. Jacobson and M. A. Delucchi, "Providing all global energy with wind, water, and solar power, Part I: Technologies, energy resources, quantities and areas of infrastructure, and materials", Energy Policy, vol. 39, p. 1154, 2011.

[4] J. A. Turner, "A realizable renewable energy future", Science, vol. 285, p. 687, 1999.

[5] F. Ueckerdt, R. Brecha, and G. Luderer, "Analyzing major challenges of wind and solar variability in power systems", Renewable Energy, vol. 81, p. 1, 2015.

[6] H. Kanchev, D. Lu, F. Colas, V. Lazarov, and B. Francois, "Energy management and operational planning of a microgrid with a PV-based active generator for smart grid applications", IEEE Trans. Ind. Electron., vol. 58, pp. 4583-4592, 2011.

[7] B. Ramachandran, S. K. Srivastava, C. S. Edrington, and D. A. Cartes, "An intelligent auction scheme for smart grid market using a hybrid immune algorithm", IEEE Trans. Ind. Electron., vol.58, pp. 4603-4612, 2011.

[8] Q. Yang, J. A. Barria, and T. C. Green, "Communication infrastructures for distributed control of power distribution networks", IEEE Trans. Ind. Inf., vol. 7, pp. 316-327, 2011.

[9] P. Milan, M. Wächter, and J. Peinke, "Turbulent character of wind energy", Phys. Rev. Lett., vol. 110, p. 13, 2013.

[10] D. Heide et al., "Seasonal optimal mix of wind and solar power in a future, highly renewable Europe", Renewable Energy, vol. 35, p. 2483, 2010.

[11] Q. Morante, N. Ranaldo, A. Vaccaro, and E. Zimeo, "Pervasive grid for large-scale power systems contingency analysis", IEEE Trans. Ind. Inf., vol. 2, pp. 165-175, 2006.

[12] I. Balaguer, Q. Lei. S. Yang. U. Supatti, and F. Z. Peng, "Control for grid-connected and intentional islanding operations of distributed power generation", IEEE Trans. Ind. Electron., vol. 58, pp. 147-157, 2011.

[13] M. Liserre, T. Sauter, and J. Y. Hung, "Future energy systems: Integrating renewable energy sources into the smart power grid through industrial electronics", IEEE Ind. Electron. Mag., vol.4, pp. 18-37, 2010.

[14] E. Santacana, G. Rackliffe, L. Tang, and X. Feng, "Getting smart", IEEE Power and Energy Magazine, vol. 8, pp. 41-48, 2010.

[15] V. Calderaro, C. Hadjicostis, A. Piccolo, and P. Siano, "Failure identification in smart grids based on petri net modeling", IEEE Trans. Ind. Electron., vol.58, pp. 4613-4623, 2011.

[16] D. Bakken, A. Bose, C. Hauser, D. Whitehead, and G. Zweigle, "Smart generation and transmission with coherent real-time data", Proceedings of the IEEE, vol. 99, pp. 928-951, 2011.

[17] F. Dörfler, M. Chertkov, and F. Bullo, "Synchronization in complex oscillator networks and smart grids", Proceedings of the National Academy of Sciences, vol. 110(6), pp. 2005-2010, 2013.

[18] B. Schäfer, C. Beck, K. Aihara, D. Witthaut, and M. Timme, "Non-Gaussian power grid frequency fluctuations characterized by Lévy-stable laws and superstatistics", Nat. Energy, vol. 3, p. 119, 2018.

[19] M. Rohden, A. Sorge, M. Timme, and D. Witthaut, "Self-organized synchronization in decentralized power grids", Phys. Rev. Lett., vol. 109(6), p. 064101, 2012.

[20] B. Schäfer, M. Matthiae, M. Timme, and D. Witthaut, "Decentral smart grid control", New J. Phys., vol. 17(1), p. 015002, 2015.

[21] B. Schäfer et al., "Taming instabilities in power grid networks by decentralized control", The European Physical Journal Special Topics, vol. 225(3), pp. 569-582, 2016.

[22] C. Wang, C. Grebogi, and M. S. Baptista, "Control and prediction for blackouts caused by frequency collapse in smart grids", Chaos, vol. 26(9), p. 093119, 2016.

[23] A. Ulbig, T. S. Borsche, and G. Andersson, "Impact of low rotational inertia on power system stability and operation", IFAC Proceedings Volumes, vol. 47, p. 7290, 2014.

[24] R. Doherty et al., "An assessment of the impact of wind generation on system frequency control", IEEE Trans. Power Syst., vol. 25, p. 452, 2010.

[25] E. Weitenberg et al., "Robust decentralized secondary frequency control in power systems: Merits and trade-offs", IEEE Trans Automat Contr, vol. 10, pp. 3967-3982, 2018.

[26] E. B. T. Tchuisseu et al., "Curing Braess' paradox by secondary control in power grids", New J. Phys., vol. 20(8), p. 083005, 2018.

[27] J. W. Simpson-Porco, F. Dörfler, and F. Bullo, "Droop-controlled inverters are Kuramoto oscillators", IFAC Proceedings Volumes, vol. 45(26), pp. 264-269, 2012.

[28] H. Okuno and M. Kawakita, "Delayed feedback control of three-synchronous-generator infinite-bus system", Electrical Engineering in Japan, vol. 156(1), pp. 7-12, 2006.

[29] E. D. Dongmo, P. Colet, and P. Woafo, "Power grid enhanced resilience using proportional and derivative control with delayed feedback", Eur. Phys. J. B, vol. 90(1), p. 6, 2017.

[30] H. Taher, S. Olmi, and E. Schöll, "Enhancing power grid synchronization and stability through time delayed feedback control", Phys. Rev. E, vol. 100, p. 062306, 2019.

[31] G. Filatrella, A. H. Nielsen, and N. F. Pedersen, "Analysis of a power grid using a Kuramoto-like model", Eur. Phys. J. B, vol. 61(4), pp. 485-491, 2008.

[32] J. Giraldo, E. Mojica-Nava, and N. Quijano, "Synchronization of dynamical networks with a communication infrastructure: A smart grid application", in 52nd IEEE Conference on Decision and Control, p. 4638, 2013.

[33] K. Schmietendorf, J. Peinke, and O. Kamps, "The impact of turbulent renewable energy production on power grid stability and quality, Eur. Phys. J. B vol. 90, p. 222, 2017.

[34] H. Li and Z. Han, "Synchronization of power networks without and with communication infrastructures", in Proceedings of the 2011 IEEE International Conference on Smart Grid Communications (SmartGridComm), pp. 463-468, 2011.

[35] J. Wei, D. Kundur, T. Zourntos, and K. Butler-Purry, "A flocking-based dynamical systems paradigm for smart power system analysis", in Power and Energy Society General Meeting, 2012 IEEE, pp. 1-8, 2012.

[36] J. Baillieul and P. Antsaklis, "Control and communication challenges in networked real-time systems", Proceedings of the IEEE, vol. 95, pp. 9-28, 2007.

[37] S. V. Buldyrev, R. Parshani, G. Paul, E. Stanley, and S. Havlin, "Catastrophic cascade of failures in interdependent networks", Nature, vol. 64, p. 1025, 2010.

[38] C. H. Totz, S. Olmi, and E. Schöll, "Control of synchronization in two-layer power grids", Phys. Rev. E, vol. 102(2), p. 022311 (2020).

Stability and Control of Power Grids with Diluted Network Topology

Liudmila Tumash
CNRS, Grenoble INP, GIPSA-lab
Grenoble, France
email: liudmilatumash@gmx.de
ORCID 0000-0002-8144-5356

Simona Olmi
Istituto dei Sistemi Complessi
CNR - Consiglio Nazionale delle Ricerche
Sesto Fiorentino, Italy
email: simona.olmi@fi.isc.cnr.it
ORCID 0000-0002-8272-3493

Eckehard Schöll
Institut für Theoretische Physik
Technische Universität Berlin
Berlin, Germany
Potsdam Institute for Climate Impact Research
Potsdam, Germany
email: schoell@physik.tu-berlin.de
ORCID 0000-0002-7318-2672

***Abstract*—We consider sparse random networks of Kuramoto phase oscillators with inertia in order to investigate the dynamics emerging in high-voltage power grids. The corresponding natural frequencies are assumed to be bimodally Gaussian distributed, thus modeling the distribution of both power generators and consumers, which must be in balance. Our main focus is on the theoretical analysis of the linear stability of the frequency-synchronized state, which is necessary for the stable operation of power grids, and the control of unstable synchronous states. We demonstrate by numerical simulations that unstable frequency-synchronized states can be stabilized by feedback control. Further, we extend our study to include stochastic temporal power fluctuations and discuss the interplay of topological disorder and Gaussian white noise for various model configurations. Our results are compared with those obtained for the real power grid topology of Italy.**

***Index Terms*—power grids, sychronization, stability, control, diluted network.**

I. Introduction

The traditional way to generate power by using fossile energy has induced the risk of global warming caused by large emission of carbon dioxide gases. Nowadays, we are witnessing a drastic regime shift in the operation of power grids towards renewable energy sources, since more and more energy generating units become supplied by natural sources, such as wind parks and photovoltaic arrays, see [14], [15]. This regime shift has triggered three major issues that have to be considered to provide the sustainable operation of power grids. The first issue is *decentralization* meaning that the power system based on renewable energy sources represents a distributed network carrying many small units of energy to the consumers (unlike conventional power grids), see [2]. The second issue is a strong *spatial separation* between power generators and consumers [4], e.g., wind energy is usually produced near the sea, and solar energy is harvested in sunny areas, while power consumption is highest in industrial agglomerations in other parts of the country. Finally, the third issue is related to the strong dependence on weather conditions, which leads to the increasing fraction of *strongly fluctuating power output* [8]. Thus, the focus of this study is on power grids based on renewable energy sources characterized by sparse networks. For this reason, we have considered random Erdös-Renyi networks with low average connectivity to model the network topology underlying high voltage transmission grids [11], [13].

The Kuramoto model with inertia presented in [1] is a standard mathematical model used to study the dynamical behavior of power generators and consumers [5]–[7], [11], [12], [16]. Thus, we consider sparse random networks of Kuramoto phase oscillators with inertia to investigate the dynamics emerging in high-voltage power grids. In general, power grids tend to synchronize their frequencies to the standard ac power frequency $\Omega = 50$ Hz (or 60 Hz in some countries). We distinguish the power generated by power sources ($P^i_{source} > 0$) from the power consumed by passive machines or loads ($P^i_{cons} < 0$) by assuming the bimodal Gaussian distribution of natural frequencies of power generators and consumers with opposite picks as in [5]. Although the bimodal frequency distribution is a very important feature of the model, most of the previous studies consider either a unimodal frequency distribution [6] or δ-function shaped bimodal distributions [7]. In our work we use the bimodal Gaussian distribution of frequencies, which models consumed and generated power in a more realistic way.

For the stable operation of power grids the produced power must be equal to the consumed power, which implies maintaining a synchronous state of the entire network. We provide the theoretical analysis of the linear stability of the frequency-synchronized state, and the control of unstable synchronous states that are usually characterized by large differences of initial phases. The stability criteria were derived based on the properties of the initial phase differences of the oscillators. We demonstrate by numerical simulations that unstable frequency-synchronized states can be stabilized by feedback control if the coupling between oscillators is strong enough. Additionally, we include stochastic temporal power fluctuations by adding Gaussian white noise to the Kuramoto model and discuss how does it influence the frequency synchronization.

The present extended abstract is structured as follows. In Sec. II, the Kuramoto model with inertia, the network topology and natural frequency distribution are presented. In Sec. III we characterize frequency synchronization and discuss the conditions under which this state occurs, as well as we apply a control method to stabilize the unstable frequency-synchronized solutions. The results and main contributions of this study are discussed in Sect. IV.

II. Model

We consider a system consisting of a population of $i = 1, \ldots, N$ coupled Kuramoto oscillators with inertia that reads

$$m\ddot{\theta}_i + \dot{\theta}_i = \Omega_i + \frac{K}{N_i}\sum_{j=1}^{N} A_{ij}\sin\left(\theta_j - \theta_i\right), \qquad (1)$$

where θ_i and $\dot{\theta}_i$ are the instantaneous phase and frequency, respectively, of the oscillator i. Parameters $m > 0$ and $K > 0$ indicate the inertial mass of generators and the coupling constant of the network equivalent to the transmission line capacities between loads and generators, respectively. A is the connectivity matrix: its entries A_{ij} are one if nodes i and j are connected, and zero otherwise. N_i is the node degree of the i-th element, thus denoting the number of the links outcoming from this node. Finally, Ω_i represents the natural frequency of the oscillator i, whose value is chosen in accordance with the bimodal Gaussian distribution

$$g(\Omega) = \left[\frac{p_g}{\sqrt{2\pi}} e^{-\frac{(\Omega-\Omega_{0+})^2}{2}} + \frac{1-p_g}{\sqrt{2\pi}} e^{-\frac{(\Omega+\Omega_{0-})^2}{2}}\right]. \qquad (2)$$

III. Frequency Synchronization and Control

In particular, we aim to investigate the stability of the synchronous solution emerging in a power grid network by linearizing the state around the frequency-synchronized solution, which gives a Jacobian matrix with constant coefficients, and analyzing the eigenvalues of this linearized system. The calculation of the eigenvalue with the largest real part λ_{max} will be the main criterion for determining the synchronization stability. Stability means that the sign of the largest real part is negative.

We analyse the frequency synchronized solution for the network characterized by coupling beyond some critical value (for $K < K_c$ no frequency synchronized solution is possible). We have obtained that stable frequency synchronization occurs in systems whose initial phases are close enough, i.e., $\left|\theta_j^0 - \theta_i^0\right| < \frac{\pi}{2}$. If this condition on phases is violated, we deal with unstable solutions, which we stabilize with a control loop.

Further, we search for sets of initial phases that satisfy the frequency-synchronized solution numerically using Levenberg-Marquardt algorithm, which might find stable and unstable solutions. For example, Fig.1a),b) presents a stable frequency-synchronized solution that is obtained for almost identical initial phases, while for the case of largely varying initial phases we obtain an unstable frequency-synchronized solution as illustrated in Fig.1c),d).

We further stabilize such unstable frequency-synchronized solutions obtained for $\left|\theta_j^0 - \theta_i^0\right| \geq \frac{\pi}{2}$ by introducing a control term u_i into the original system (1)

$$\dot{\theta}_i = \omega_i$$

$$\dot{\omega}_i = \alpha\Omega_i - \alpha\omega_i + \frac{K}{N_i}\sum_{j=1}^{N} A_{ij}\sin(\theta_j - \theta_i) + u_i,$$

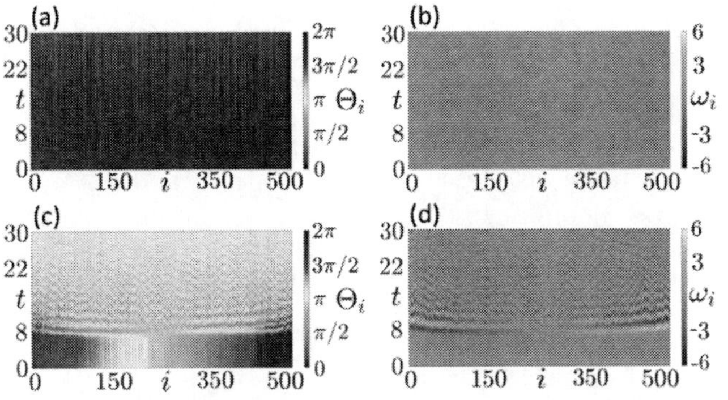

Fig. 1. Spatio-temporal evolution of phases θ_i and frequencies ω_i, which satisfy the condition for frequency-synchronized solution. Stable solution: (a) phases; (b) frequencies; parameters: $Re(\lambda_{max}) = -0.083$, $K = 10$. Unstable solution: (c) phases; (d) frequencies; parameters: $\lambda_{max} = 2.41$, $K = 70$. Other parameters: $m = 6$, $p = 0.20$ (connectivity ratio), $\Omega_0 = 2$, $N = 500$.

In particular, the control term u can be chosen as a feedback control loop such that

$$u = -\mathbf{C}\begin{pmatrix}\delta\theta \\ \delta\omega\end{pmatrix},$$

where $\mathbf{C} \in R^{N\times 2N}$ is chosen to minimize the following cost functional

$$J(u) = \int_0^{\infty} \left|\left|\begin{pmatrix}\delta\theta(t) \\ \delta\omega(t)\end{pmatrix}\right|\right|^2 + ||u(t)||^2 dt.$$

This problem is solved via the application of a *linear quadratic regulator* for each set of phases θ_i^*. Basically, the regulator chooses the time-independent matrix $\mathbf{C}$ such that the eigenvalues for the closed-loop system are non-positive when solving the eigenvalue problem for the Jacobian. Thus, the frequency-synchronized solution is stabilized for each particular set of chosen phases θ_i^*, and, regardless of the initial phase differences $\left|\theta_j^0 - \theta_i^0\right|$, we are always able to obtain a stable solution if $K > K_c$.

An example is illustrated in Fig.2, where plots a) and b) represent phase evolution without control. We see that these phases lose synchrony with respect to frequencies, since their initial phases do not have close values. However, if a control action is performed, the frequency-synchronization is stabilized as it is shown in Fig.2c) and d).

IV. Conclusions

In conclusion, we have considered sparse networks of Kuramoto oscillators with inertia to investigate the optimal conditions for the emergence of synchronization in power grids. Going beyond previous work [11] devoted to this type of networks, we have provided a linear stability analysis of the frequency-synchronized solution that is necessary for the stable operation of power grids. We have derived the stability criteria, based on the initial phase differences of the oscillators, and have estimated the critical coupling strength K_c above which a frequency-synchronized solution is possible in the deterministic system. For sufficiently large coupling we have

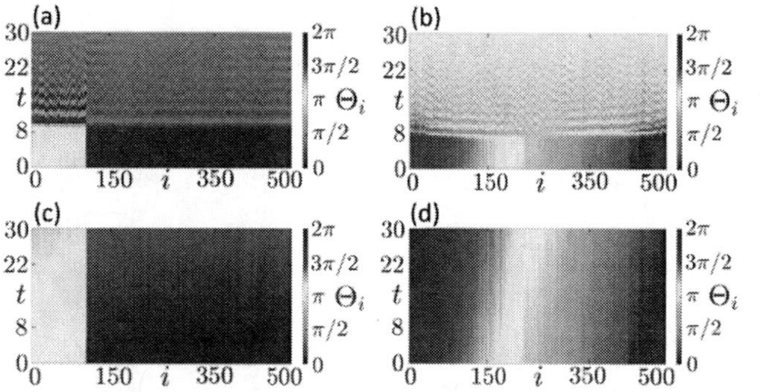

Fig. 2. Spatio-temporal evolution of phases θ_i without (top panel) and with control (bottom panel). Left column: $K = 50$ and initial phases $\theta_1^* = ... = \theta_{150}^* = \pi$, $\theta_{151}^* = ... = \theta_{500}^* = 0$, (a) control off, $\lambda_{max} = 2.802$; (c) control on, $\lambda_{ctrl} = -0.759$. Right column: $K = 70$ and uniformly distributed initial phases, (b) control off, $\lambda_{max} = 2.41$; (d) control on, $\lambda_{ctrl} = -0.823$. Other parameters: as on Fig. 1.

also found unstable solutions that are usually characterized by large differences of initial phases. A similar stability analysis was performed by Mirollo et al. [3] and by Delabays et al. [9] for networks of classical Kuramoto oscillators (without inertia) with different topologies (fully coupled networks and planar graphs respectively). Here we used the stability analysis to characterize unstable synchronous states in diluted networks, which we subsequently stabilize by a control loop. It turns out that our linear feedback control scheme is very efficient in stabilizing unstable frequency-synchronized solutions for arbitrary initial phases, and all $K > K_c$.

Furthermore we have investigated diluted networks with stochastic dynamics due to temporally fluctuating power in order to infer the similarities and differences occurring in the transition to synchronization with respect to the deterministic case. We have added a simple noise term, i.e., Gaussian white noise, rather than correlated noise or intermittent noise, in order to gain insight into the general role played by noise in power systems. Previously, the transition to synchronization has been investigated mainly in deterministic systems [6] or in globally coupled networks [11]. On the other hand, when stochastic systems with Gaussian white noise were considered [10], the focus has not been on the synchronization transition, thus neglecting possible consequences of hysteresis in power systems. In particular, here we have observed that for synthetic diluted networks (independently of the frequency distribution), intermediate noise intensities might play a constructive role in lowering the critical coupling value required to reach (almost complete) frequency synchronization, since noise suppresses intermediate states and reduces the hysteretic region.

Future perspectives of this work might be aimed at a deeper understanding of the applicability of the control scheme within noisy systems, for non-Gaussian noise and realistic topologies.

ACKOWLEDGMENT

We acknowledge support from the Deutsche Forschungsgemeinschaft (DFG) in the framework of the SFB 910, Projektnummer 163436311.

REFERENCES

[1] H. A. Tanaka, A. J. Lichtenberg, and S. Oishi, "First order phase transition resulting from finite inertia in coupled oscillator systems", Phys. Rev. Lett., vol. 78, pp. 2104-2107, 1997.
[2] T. Ackermann, G. Andersson, and L. Söder, "Distributed generation: a definition", Elec. Pow. Sys. Res., vol. 57, pp. 195-204, 2001.
[3] R. E. Mirollo and S. H. Strogatz, "The spectrum of the locked state for the Kuramoto model of coupled oscillators", Phys. D: Nonlin. Phenom., vol. 205, pp. 249-266, 2005.
[4] J. C. Smith, M. R. Milligan, E. A. DeMeo, and B. Parsons, "Utility wind integration and operating impact state of the art", IEEE transactions on power systems, vol. 22, pp. 900-908, 2007.
[5] G. Filatrella, A. H. Nielsen, and N. F. Pedersen, "Analysis of a power grid using a Kuramoto-like model", Eur. Phys. J. B, vol. 61, pp. 485-491, 2008.
[6] S. Olmi, A. Navas, S. Boccaletti, and A. Torcini, "Hysteretic transitions in the Kuramoto model with inertia", Phys. Rev. E, vol. 90, p. 042905, 2014.
[7] M. Rohden, A. Sorge, D. Witthaut, and M. Timme, "Impact of network topology on synchrony of oscillatory power grids", Chaos, vol. 24, p. 013123, 2014.
[8] M. Anvari, M. Wächter, and J. Peinke, "Phase locking of wind turbines leads to intermittent power production", Europhys. Lett., vol. 116, p. 60009, 2017.
[9] R. Delabays, T. Coletta, and P. Jacquod, "Multistability of phase-locking in equal-frequency Kuramoto models on planar graphs", Journal of Mathematical Physics, vol. 58, p. 032703, 2017.
[10] B. Schäfer, M. Matthiae, X. Zhang, M. Rohden, M. Timme, and D. Witthaut, "Escape routes, weak links, and desynchronization in fluctuation-driven networks", Phys. Rev. E, vol. 95, p. 060203, 2017.
[11] L. Tumash, S. Olmi, and E. Schöll, "Effect of disorder and noise in shaping the dynamics of power grids", Europhys. Lett., vol. 123, p. 20001, 2018.
[12] H. Taher, S. Olmi, and E. Schöll, "Enhancing power grid synchronization and stability through time delayed feedback control", Phys. Rev. E, vol. 100, p. 062306, 2019.
[13] L. Tumash, S. Olmi, and E. Schöll, "Stability and control of power grids with diluted network topology", Chaos, vol. 29, p. 123105, 2019.
[14] D.-N. Ciupageanu, G. Lazaroiu, and L. Barelli, "Wind energy integration: Variability analysis and power system impact assessment", Energy, vol. 185, pp. 1183-1196, 2019.
[15] L. Barelli et al., "Adaptive voltage control of islanded RES-based residential microgrid with integrated flywheel/battery hybrid energy storage system", 22nd European Conference on Power Electronics and Applications, pp. 1-10, 2020.
[16] C. H. Totz, S. Olmi, and E. Schöll, "Control of synchronization in two-layer power grids", Phys. Rev. E, vol. 102, p. 022311, 2020.

Dynamics of Momentary Reserves under Contingency: Observations from Numerical Experiments

Kosisochukwu Pal Nnoli
Dept. of Computer Science and Electrical Engineering
Jacobs University, 28759 Bremen
Bremen, Germany
email: k.nnoli@jacobs-university.de

Stefan Kettemann
Dept. of Physics and Earth Science and Dept. of Computer Science
Jacobs University, 28759 Bremen
Bremen, Germany
Division of Advanced Materials Science, Pohang 790-784
South Korea
email: s.kettemann@jacobs-university.de

***Abstract*—This paper presents studies and investigations on the dynamics of momentary reserves in electrical power systems under contingency. Momentary reserve through the machine's inertia serves the purpose of primary frequency control and prevents voltage collapse in the case of reactive power reserves. A simulation was performed on a realistic Nigerian 330 kV transmission network in PowerFactory software to study and investigate the mechanism of these reserve functions on the network buses as an inertia active power control method. Moreover, we investigated the influence of geodesic increment of momentary reserve on the decay of disturbances. The results indicated that the momentary reserve by inertia alone reduces the frequency deviation from its nominal value, delays the transmission of disturbances and enhances the damping of oscillations by reducing the final frequency settling time at the buses under contingency. This numerical experiment also suggests the optimal placement of the momentary reserves in the grid in order to improve system stability against power outage disturbances.**

***Index Terms*—momentary reserve; system disturbance; inertia control, oscillations damping; frequency stability.**

I. Introduction

The active and reactive power reserves of synchronous generators or Battery Energy Storage Systems (BESS) are the keys to a successful system control in power systems. For each power generator in the grid, power reserve represents the total amount of power remaining after the supply of system loads and losses. Of course, this definition does not extend to exceeding the power capability curve of the generators. These reserves are particularly referred to as spinning reserves in synchronous machines more than in other kinds of power generators, like the wind and solar generators using BESS [1]. These reserves can be used for both or either primary frequency control, secondary control and tertiary control [2]–[4]. Again, not all of these remaining power reserves from generators are assigned for primary control function alone, we classify the ones momentarily made available through the generators' droop functions specifically for few seconds primary frequency and active power control as momentary reserves of the generators. A lot about system stability and reliability rest on the grid's momentary reserve.

The control and response of generators to network disturbances or contingencies depend heavily on the kind and magnitude of the disturbance and amount of momentary reserves available for use in the primary frequency control or in voltage security. Since primary control requires fast control action to be taken within few seconds of contingency, momentary reserves and their placement play important roles in the dynamics of the disturbance from the event or fault location to the rest of the electrical power network. In this paper, we investigate how the dynamics of a disturbance are influenced by momentary reserves first at the contingent node, its neighbour nodes and other nodes located far away from the fault location. We will see whether a disturbance is damped as it travels across a grid and whether it could be contained on fewer nodes (i.e., localized) based on the function of the grid's momentary reserves. These investigations are carried out in the DigSILENT PowerFactory software [5], using the Nigerian 330 kV grid as a case study.

In this paper, we will start with the description, modeling and simulation of our test transmission network in Section II. In Section III, we will investigate the dynamics of the momentary reserve on the test network and conclude with our findings and recommendation in Section IV.

II. Modeling and Simulation of the Nigerian Transmission Network Model

To understand our case study system and its parameter interactions, we will describe the components that make up the network. The Nigerian 330 kV transmission grid consists of $N_S = 71$ substations/nodes, $N_L = 81$ over-head transmission lines (from an alloy of aluminium and steel) with an average length of 92 km, each with a limiting current of 1320 A. The grid is comprised of 107 less-decommissioned units of generators, accounting for the present 29 power stations. The active power capacity of the Nigerian grid is about $13,208MW$ as of 2020 [6]. There are other lower voltage networks including the 132 kV and 33 kV sub-transmission networks. For household utilities, there are 11 kV and 0.415 kV 3-phase distribution networks. The Nigerian network operates at $v_0 =$ 50 Hz frequency and can be described as a grid where most of the nodes are connected to one another in a ring form [7], as seen in the diagram shown in Figure 1.

To control the voltage outputs of the generators through their excitation control, the simplified excitation Automatic Voltage Regulator (AVR) model is used [8]. Other controllers also include Power System Stabilizers (PSS2A-model) tasked with enhancing the damping of the entire power system's oscillations through excitation control.

The input signal to the PSS2A controller is the derivative of generator's rotor speed injected to the AVR through the excitation system. This injection works to terminate the phase-lag between the voltage reference and the windings' torque of the generator [9]. For the synchronous machine model, the choice of the model is influenced by the IEEE guide in [10]. The speed governor in the model is the Turbine Governor (TGOV) model used to maintain the frequency operational limits according to swing equation [11]. Here, the swing equation describes the torque balance between the mechanical torque T_t in N.m. of each synchronous machine's turbine and the electromagnetic torque T_e in N.m. as governed by the differential equation given as [2], [3], [12],

$$J_i \frac{d\omega_i^{\Theta}}{dt} + D_{r_i}\omega_i^{\Theta} = T_t - T_e - D_{r_i}\omega_0, \tag{1}$$

where $J_i = \frac{2H_i}{\omega_0^2} S_i$ is the combined moment of inertia of the generator and turbine in kg.m^2 with H_i the generator inertia constant in seconds and S_i the generator apparent power in Volt-Amperes (VA). D_{r_i} represents the rotational loss due to generator rotor windings for each ith generator in N.m.s and i denotes the index of power generators in the grid. Here, t is time in seconds and ω_i^{Θ} is the angular velocity of the rotor in electrical rad/s with ω_0 as its rated synchronous value in electrical rad/s. If we assume that a change in the rotor's angular velocity $(\omega_i^{\Theta} - \omega_0)$ is a derivative of its angular position δ in electrical radians with respect to its rotating setpoint, δ_0 at t = 0 given as

$$\omega_i^{\Theta} - \omega_0 = \frac{d\delta_i}{dt}, \tag{2}$$

then, with respect to time, the derivative of ω_i^{Θ} would give

$$\frac{d\omega_i^{\Theta}}{dt} = \frac{d}{dt}\left(\frac{d\delta_i}{dt}\right) + \frac{d\omega_0}{dt}, \tag{3}$$

where ω_0 is the constant rated synchronous value whose derivative with respect to time gives zero (i.e., $\frac{d\omega_0}{dt} = 0$), (3) becomes,

$$\frac{d\omega_i^{\Theta}}{dt} = \frac{d^2\delta_i}{dt^2}. \tag{4}$$

In practice, ω_0 is related to the grid frequency (v_o) by $2\pi v_o$, where v_o is 50 Hz in the Nigerian power grid. If we represent the net mechanical shaft torque at grid frequency to be $T_m = T_t - D_{r_i}\omega_0$, substituting (4) into (1), we then have,

$$J_i \frac{d^2\delta_i}{dt^2} + D_{r_i}\left(\frac{d\delta_i}{dt}\right) = T_m - T_e. \tag{5}$$

Here, we assumed that the network perturbation effected on the rotors from the fault location is small. Multiplying both sides of (5) by the rated speed (ω_0) in order to ensure that we maintain a synchronous 50 Hz revolution throughout the system, balancing the power, we have

$$J_i\omega_0 \frac{d^2\delta_i}{dt^2} + D_{r_i}\omega_0\left(\frac{d\delta_i}{dt}\right) = T_m\omega_0 - T_e\omega_0. \tag{6}$$

As power $P = T\omega$, the right side of (6) can now be written as

$$J_i\omega_0 \frac{d^2\delta_i}{dt^2} + D_{r_i}\omega_0\left(\frac{d\delta_i}{dt}\right) = P_m - P_e, \tag{7}$$

where P_m is the turbine's mechanical power and P_e is the generator's air-gap electrical power. If we represent the rotor angular momentum at rated speed with M_i (i.e., $M_i = J_i\omega_0 = \frac{2H_i}{\omega_0} S_i$) and also represent the damping coefficient at rated synchronous speed with D_i (i.e., $D_i = D_{r_i}\omega_0$), the swing equation can then be re-written in many forms as,

$$M_i \frac{d^2\delta_i}{dt^2} + D_i\left(\frac{d\delta_i}{dt}\right) = P_m - P_e, \tag{8}$$

and also as [13], [14],

$$\frac{2H_i}{\omega_o} S_i \frac{d^2\delta_i}{dt^2} + D_i \frac{d\delta_i}{dt} = P_i + \sum_{j=1}^{N_S} W_{ij}\sin(\delta_j - \delta_i). \tag{9}$$

Where P_i is the power in the grid's i^{th} node, N_S is the number of nodes/buses and W_{ij} is the power capacity in Watt of the transmission lines and it is dependent on the network voltage with $\sin(\delta_j - \delta_i)$ modeling the dependence of their phase differences which informs the direction of powerflow and the transmission of disturbances in the case of contingencies. We performed load-flow calculations using the Newton-Raphson method and electromechanical simulations in DigSILENT PowerFactory software, as documented in [15]. Here, we report the results applying these simulations to study the effect of momentary reserves on system dynamics and its contribution to the overall system stability.

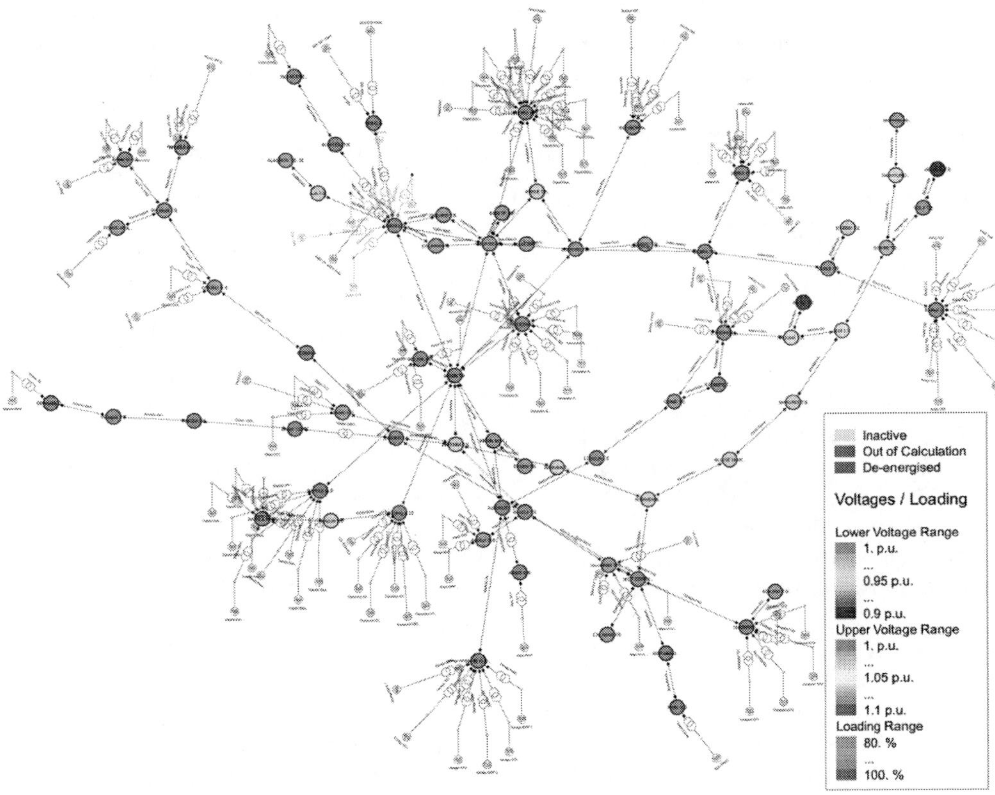

Figure 1. The Nigerian 330 kV Electrical Power Grid.

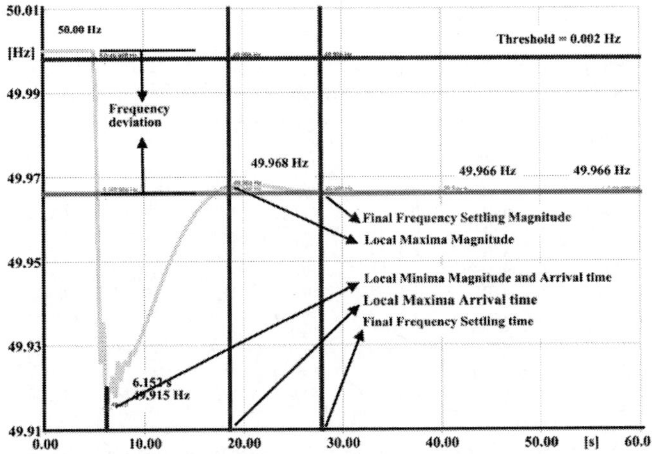

Figure 2. Green Line: Frequency as a function of time at a node where a Synchronous Machine (SM) event occurred at $t = 5$s.

III. Dynamics of Momentary Reserve

For a power system to be in a balanced state, the total power generated (P) has to be equal to the sum of the total load and the transmission losses (P^{x}). This means that for the system to always remain balanced, any change in the load demand and/ losses would require a balancing change in the generated power P. Therefore, the condition $\Delta P >> \Delta P^{x}$ should always be fulfilled to ensure the security and integrity of power grids. These dynamic changes result in a permanently changing frequency according to the speed-droop characteristics of power turbines, given by [3]

$$\frac{\Delta \nu_i}{\nu_0} = -\sigma_i \frac{\Delta P_i}{P_i^{x}}, \tag{10}$$

where ν_0 is the nominal frequency in Hertz and σ_i is the local droop of the generation characteristics at the ith generator node, $\sigma_i = P_i^{x}/P_i$. Thus, any change in frequency resulting from a change in load demand or losses requests a generator response. In this way, if the node is a synchronous generator node, its inertia is expected to slow down the transmission of disturbances to other nodes connected to it, while attempting to damp the frequency oscillations with its momentary reserve according to its droop σ_i function. However, its ability to do so depends on the amount of momentary reserve that the generator can supply, according to its dead zone speed-droop characteristics setting. Since it is not feasible to inject power equally at every node, we aim to understand the dynamics of these reserves to see how an available momentary reserve at one node influences the frequency dynamics at other nodes as function of their distance from the fault location. A better understanding of this influence in the realistic Nigerian grid model would be advantageous for its optimal placement and could improve primary frequency control in real power grids.

To show the influence of geographical distribution of inertia with available momentary reserve, we chose the Nigerian transmission grid as the case study grid. As reported in Section II, we modelled this network in PowerFactory software and calculated the load flow of the network. We considered the voltage dependency of grid loads, and the active power control according to the grid inertia. The effect of provision of momentary reserves by system inertia on the frequency stability of the electrical power networks is studied by varying the aggregated inertia constant H_{agg} of the entire grid, as defined by [16],

$$H_{\text{agg}} = \frac{\sum_{i=1}^{N_S} H_i S_i}{S_n}, \tag{11}$$

where S_i and H_i are the rated apparent power and inertia constant of the ith bus, respectively, $S_n = \sum_{i=1}^{N_S} S_i$ is the total rated power in the grid. Note that H_i on non-generator buses are set to zero. The H_{agg} was varied in the entire grid by multiplying the S_i with a common factor. Note that the Nigerian transmission network does not comprise tie-line connections and power transfer [6], [17]. The grid frequency is operated uniformly at 50 Hz across the entire network.

Here, we choose 11 buses for the investigation. The fault location bus 24, two buses at the same geodesic distance (unweighted), $r = 2$ from fault location with no inertia (i.e., buses 8 and 10) and three buses with inertia (i.e., buses 22, 55, and 57). We again choose two buses at the same geodesic distance (unweighted), $r = 7$ with no inertia (i.e., buses 7 and 30) and three buses with inertia (i.e., buses 3, 28, and 69). The reason is that we want to investigate the effect of momentary reserve both in the vicinity of the reserve (i.e., near to the fault location) and at far distances from the injection node. We also want to understand how momentary reserve tentatively contributes to damping of disturbances as they propagate along the network. A reference to the network buses numbers is in [18].

Keeping all system parameters, generations, losses and loads constant under undisturbed operations, a synchronous machine outage event (disturbance/contingency) is induced at bus 24 at exactly $t = 5$s of the 90s transient electromechanical stability simulation time frame with a 200ms switching, enabling the observations on the buses in PowerFactory power simulation software. The change in frequency propagated across the grid is related to the change of power in (10). Figure 2 shows the frequency dynamics at a network bus. Here, the red line marks the frequency magnitude at final settling time. The black arrows point to the local maxima or minima of the frequency magnitude and arrival times. From this nodal points described in Figure 2, we observe at each study case node, the frequency's Time of Arrival (ToA), which is defined as the time when the frequency deviation first reaches a small threshhold of $\delta v = 0.002$ Hz, as defined in more detail in [18]. Furthermore we observe the time of the first maximum (maxima_t) of the transient and its magnitude ($\text{maxima}_{\text{mag}}$), the time of the first minimum (minima_t) and its magnitude ($\text{minima}_{\text{mag}}$). We also record the final Frequency's Settling time (FS_t), its magnitude (FS_{mag}) and frequency deviation (Dev_{mag}) from the nominal 50Hz, for each of these nodes. In Tables I-V, we show the observations in milliseconds for the case study nodes, with static network power flow of the Nigerian transmission grid.

TABLE I. NODAL OBSERVATIONS WITH LARGE DISTURBANCE AND NO RESERVE AT FAULT LOCATION GIVEN THAT $H_{\text{agg}} = 2$s

Bus	r	ToA (s)	minima_t (s)	$\text{minima}_{\text{mag}}$ (Hz)	FS_t (s)	FS_{mag} (Hz)	Dev_{mag} (Hz)
24	0	5.012	6.522	49.596	27.200	49.829	0.171
8	2	5.012	6.522	49.596	27.201	49.829	0.171
10	2	5.012	6.522	49.596	27.200	49.829	0.171
22	2	5.013	6.282	49.596	27.195	49.829	0.171
55	2	5.013	6.532	49.596	27.201	49.829	0.171
57	2	5.012	6.532	49.597	27.196	49.829	0.171
7	7	5.013	6.372	49.598	27.205	49.829	0.171
30	7	5.013	6.322	49.600	27.206	49.829	0.171
3	7	5.013	6.422	49.590	27.185	49.829	0.171
28	7	5.013	6.942	49.596	27.184	49.829	0.171
69	7	5.013	6.352	49.592	27.186	49.829	0.171

TABLE II. NODAL OBSERVATIONS WITH LARGE DISTURBANCE AND LARGE RESERVE AT FAULT LOCATION GIVEN THAT $H_{\text{agg}} = 2$s

Bus	r	ToA (s)	minima_t (s)	$\text{minima}_{\text{mag}}$ (Hz)	FS_t (s)	FS_{mag} (Hz)	Dev_{mag} (Hz)
24	0	5.014	6.132	49.920	25.700	49.967	0.033
8	2	5.014	6.112	49.920	25.691	49.967	0.033
10	2	5.014	6.102	49.920	25.690	49.967	0.033
22	2	5.015	6.282	49.919	25.687	49.967	0.033
55	2	5.014	6.082	49.920	25.691	49.967	0.033
57	2	5.015	6.192	49.919	25.688	49.967	0.033
7	7	5.016	6.292	49.920	25.696	49.967	0.033
30	7	5.016	6.372	49.920	25.698	49.967	0.033
3	7	5.017	6.252	49.918	25.679	49.967	0.033
28	7	5.017	6.142	49.916	25.678	49.967	0.033
69	7	5.017	6.182	49.918	25.680	49.967	0.033

TABLE III. NODAL OBSERVATIONS WITH LARGE DISTURBANCE AND LARGE RESERVE AT FAULT LOCATION GIVEN THAT $H_{\text{agg}} = 6$s

Bus	r	ToA (s)	minima_t (s)	$\text{minima}_{\text{mag}}$ (Hz)	FS_t (s)	FS_{mag} (Hz)	Dev_{mag} (Hz)
24	0	5.015	9.312	49.928	29.304	49.967	0.033
8	2	5.015	9.262	49.928	29.304	49.967	0.033
10	2	5.015	9.272	49.928	29.304	49.967	0.033
22	2	5.017	8.352	49.928	29.302	49.967	0.033
55	2	5.015	9.442	49.928	29.305	49.967	0.033
57	2	5.017	9.142	49.928	29.302	49.967	0.033
7	7	5.019	9.182	49.928	29.308	49.967	0.033
30	7	5.020	8.932	49.928	29.309	49.967	0.033
3	7	5.028	8.672	49.927	29.296	49.967	0.033
28	7	5.074	8.602	49.927	29.295	49.967	0.033
69	7	5.029	8.652	49.927	29.296	49.967	0.033

In Table I and with network H_{agg} at 2s, a typical behaviour of high renewable energy source injected grids, we observe a high frequency deviation (i.e., 0.171 Hz) induced by the large disturbance (of 1320MW power magnitude outage) at bus 24 fault location. We also observe the disturbance arrival at the fault location first and at the same time as most of its nearest neighbours at $r = 2$, but arrived a little later at the distant buses 7, 30, 28 and 69. The frequency dip described by the $\text{minima}_{\text{mag}}$ shows the lowest magnitude of frequency deviation before the actions of generator governor-turbine.

At $H_{\text{agg}} = 2$s and in comparison with Table I, Table II shows the observations when the fault location is injected with

TABLE IV. NODAL OBSERVATIONS WITH LARGE DISTURBANCE AND LARGE RESERVE AT FAULT LOCATION AND AN INCREASED RESERVE AT BUS 22 GIVEN THAT $H_{agg} = 2$s

Bus	r	ToA (s)	$minima_t$ (s)	$minima_{mag}$ (Hz)	FS_t (s)	FS_{mag} (Hz)	Dev_{mag} (Hz)
24	0	5.014	6.612	49.920	27.645	49.967	0.033
8	2	5.014	6.122	49.921	27.647	49.967	0.033
10	2	5.014	6.612	49.920	27.645	49.967	0.033
22	2	5.015	6.352	49.919	27.641	49.967	0.033
55	2	5.014	6.582	49.920	27.647	49.967	0.033
57	2	5.015	6.152	49.921	27.642	49.967	0.033
7	7	5.016	6.342	49.921	27.653	49.967	0.033
30	7	5.016	6.452	49.921	27.655	49.967	0.033
3	7	5.017	6.282	49.919	27.630	49.967	0.033
28	7	5.017	6.142	49.917	27.629	49.967	0.033
69	7	5.017	6.252	49.919	27.631	49.967	0.033

TABLE V. NODAL OBSERVATIONS WITH LARGE DISTURBANCE AND LARGE RESERVE AT FAULT LOCATION AND WITH A NEWLY INSTALLED RESERVE AT BUS 7 GIVEN THAT $H_{agg} = 2$s

Bus	r	ToA (s)	$minima_t$ (s)	$minima_{mag}$ (Hz)	FS_t (s)	FS_{mag} (Hz)	Dev_{mag} (Hz)
24	0	5.014	6.662	49.920	39.118	49.967	0.033
8	2	5.014	6.632	49.920	39.116	49.967	0.033
10	2	5.014	6.662	49.920	39.118	49.967	0.033
22	2	5.015	6.332	49.918	39.127	49.967	0.033
55	2	5.014	6.632	49.920	39.115	49.967	0.033
57	2	5.015	6.202	49.919	39.124	49.967	0.033
7	7	5.016	6.602	49.920	39.102	49.967	0.033
30	7	5.016	6.492	49.920	39.097	49.967	0.033
3	7	5.017	6.292	49.919	39.149	49.967	0.033
28	7	5.017	6.152	49.917	39.151	49.967	0.033
69	7	5.017	6.272	49.919	39.148	49.967	0.033

inertia and large reserve. We observe that there are delays in the frequency ToA at the buses, a reduction in $minima_t$ which corresponds to a reduced frequency dip (i.e., $minima_{mag}$), a shorter final frequency settling time and smaller magnitude and hence, a decrease in the frequency deviation from the nominal value (0.033 Hz).

In Table III, we kept the network parameters constant and only increased the grid inertia, $H_{agg} = 6$s using (11) without changing the active power injections at any network node. We did not observe any further decrement in the frequency deviation or any increment in the frequency final settling magnitude, rather we observe a delay in its time of arrival at the buses with a corresponding increase in its dip and final settling time. This suggests that increasing the H_{agg} without a corresponding increase in the reserve does not improve the frequency dynamics during contingencies, rather it increases the arrival time of the disturbance while reducing the frequency dip across the network (i.e., improved $minima_{mag}$). This further delay in ToA also increases the frequency final settling time.

Table IV shows our nodal observations on the case study buses when we injected more power reserve prior to contingency at another bus with $r = 2$, from the fault location and keeping all other system operation parameters constant from observations in Table II. Here, the generator's power and reserve at the fault location are higher in magnitude than the injected ones. We observe that the frequency final settling magnitude did not increase, the ToA at the buses remained the same but there is an observable reduction in the frequency dip at some nodes.

To investigate the effect of the same reserve at a bus geometrically farther away from the fault location but with a higher degree of connectivity, which defines the number of edges connected to it, we removed the reserve at bus 22 with a node degree of 2 and installed it at another bus with $r = 7$ but with a node degree of 5 and keeping all other system operation parameters constant from our observations in Table II. The result of this new installation in Table V compared with Table IV shows no observable change in the frequency ToA at the buses and in the frequency final settling magnitude. Rather, we observe a delay in frequency dip time (i.e., $minima_t$) with a further delay in the final frequency settling time at the buses.

In summary, we observed that an increase in the nodal momentary reserve generally delays the travel and arrival of disturbances in a power grid at contingencies. In particular, it improves the frequency magnitude at the local minima point and reduces its final settling time. We found out that the optimal placement of momentary reserve is at the point of contingency as it contributes more in the damping of disturbance across the network more than in any other place. Since we may not always be able to predict a fault location, the optimal solution would be to place momentary reserve at all buses where resources allow. In this way, the power system could quickly recover most contingencies within few seconds after their occurrences. Again, we observed that injecting momentary reserve at a bus with high connectivity does not improve the frequency dip time (i.e., $minima_t$) and final settling time at the buses if the bus is geometrically farther away from the fault location. Hence, the farther away the reserve is from the fault location, the more time it would take for the frequency to stabilize at the buses.

Also, we conclude that increasing the grid inertia without a corresponding increase in the magnitude of the reserve could only delay the travel and arrival of disturbances in electrical network but does not reduce the frequency deviation from the nominal value. This result would be important to the Transmission System Operators (TSOs) when injecting virtual inertia in the energy transition to renewable schemes.

IV. CONCLUSION

In this paper, we have studied the dynamics of power system reserves. By realistic numerical experiments, we explored how they contribute to the damping of system oscillations at contingencies and thereby work to restore the grid frequency to its nominal value. We have shown that the optimal placement of momentary reserve would be at the fault location, particularly in the case of generator outage events at plant stations. Since this situation could not always be predicted, we suggest placement of reserves in all the nodes where resources permit as this would improve the overall final frequency settling time, frequency dips, and reduce overall frequency deviation from the nominal value, thereby contributing to primary frequency control and reducing the amount of secondary control needed.

Furthermore, our detailed work on frequency dynamics and spread of disturbances in many case study grids can be found in [18].

ACKNOWLEDGMENT

We gratefully acknowledge the support of Bundesministerium für Bildung und Forschung (BMBF) CoNDyNet-2, FK. 03EK3055D.

REFERENCES

[1] K. Shi, H. Ye, W. Song, and G. Zhou, "Virtual inertia control strategy in microgrid based on virtual synchronous generator technology," *IEEE Access*, vol. 6, pp. 27 949–27 957, 2018.

[2] P. Kundur, *Power System Stability and Control*. McGraw-Hill, New York, 1994.

[3] J. Machowski, J. Bialek, and J. R. Bumby, *Power System Dynamics: Stability and Control*. Wiley, 2008.

[4] P. Anderson and A. Fouad, "Power system control and stability," *Iowa State University Press, Ames, Iowa*, vol. 1, 1977.

[5] DigSILENT GmbH, *PowerFactory version 2020 Software Manual*, Heinrich-Hertz-Straße 9, 72810, Gomaringen, Germany, 2020.

[6] K. P. Nnoli, "Implementation of a dynamic network model of the Nigerian transmission grid for investigations on power system stability," *https://doi.org/10.31224/osf.io/r82zn*, 2019.

[7] National Control Center Osogbo, "Transmission data update of the transmission company of Nigeria," *Field Research at National Transmission Company of Nigeria in March*, 2019.

[8] ENTSO-E, *Documentation on Controller Test in Test Grid Configurations*. Entso-E, 2013.

[9] G. Rogers, "Demystifying power system oscillations," *IEEE Computer Applications in Power*, vol. 9, no. 3, pp. 30–35, 1996.

[10] IEEE Power Engineering Society, *IEEE Guide for Synchronous Generator Modelling Practice and Applications in Power System Stability Analysis*. IEEE Press, 2002.

[11] IEEE Power and Energy Society, *Dynamic Models for Turbine-Governor in Power System Studies*. IEEE Press, 2013.

[12] A. A. Sallam and O. P. Malik, *Power System Stability: Modelling, Analysis and Control*. IET Power and Energy Series 76, 2015.

[13] A. R. Bergen and D. J. Hill, "A structure preserving model for power system stability analysis," *IEEE Transactions on Power Apparatus and Systems*, vol. PAS-100, no. 1, pp. 25–35, 1981.

[14] D. Manik, M. Rohden, H. Ronellenfitsch, X. Zhang, S. Hallerberg, D. Witthaut, and M. Timme, "Network susceptibilities: Theory and applications," *Physical Review E 95, 012319*, 2017.

[15] K.P. Nnoli and S. Kettemann, "Supplementary material in DataPort," *[IEEE DataPort] https://dx.doi.org/10.21227/pjpt-nk47*, 2021.

[16] A. Ulbig, T. S. Borsche, and G. Anderson, "Impact of low rotational inertia on power system stability and control," vol. 14, no. 3, pp. 7290–7297, 2014.

[17] Nigerian Electricity Regulatory Commission (NERC), *The Grid Code for the Nigerian Electricity Transmission System*. NERC Press, 2014.

[18] K.P. Nnoli and S. Kettemann, "Spreading of disturbances in realistic models of transmission grids: Dependence on topology, inertia and heterogeneity," *[engrarxiv] https://dx.doi.org/10.31224/osf.io/c8awt*, 2021.

Data Analysis of Frequency Fluctuations in the Balearic Grid Before and After Coal Closure

María Martínez-Barbeito
Instituto de Física Interdisciplinar y Sistemas Complejos, IFISC (CSIC-UIB)
Palma de Mallorca, Spain
email: maria@ifisc.uib-csic.es

Damià Gomila
Instituto de Física Interdisciplinar y Sistemas Complejos, IFISC (CSIC-UIB)
Palma de Mallorca, Spain
email: damia@ifisc.uib-csic.es

Pere Colet
Instituto de Física Interdisciplinar y Sistemas Complejos, IFISC (CSIC-UIB)
Palma de Mallorca, Spain
email: pere@ifisc.uib-csic.es

Abstract—**In 2019, the most polluting power station in the Balearic Islands was partially closed down, marking the end of coal as the main energy source in the territory. In this work, we analyze the differences in the statistics of fluctuations of the electrical frequency before and after the closure.**

Keywords—frequency fluctuations; power grid analysis; frequency-dependent control; coal generation; decarbonization.

I. Introduction

Stable power grid operation is based on the continuous balance between supply and demand. This balance is not trivial due to the lack of large-scale storage capacity and the intrinsic fluctuations of (part of) the demand. If demand exceeds generation, the grid frequency reduces, while if generation exceeds demand, it increases. Thus, studying frequency fluctuations is a good proxy to analyze the power grid stability.

Nowadays, balance is achieved by adapting in real time the generated power to the demand, which can only be achieved using controllable energy sources. Besides being controllable, conventional power plants play a key role in grid stability. They provide primary and secondary control, which compensate frequency deviations from the reference frequency, and they incorporate large inertia to the grid, which damps fast fluctuations.

The need to urgently address the effects of climate change and the dependency of the energy sector on hydrocarbon resources is accelerating the transition towards sustainable and renewable energies. A first step in this transition implies the progressive closure of the most contaminating power plants, such as those based on coal, whose role is taken over by cleaner conventional energy sources, e.g., natural gas. Subsequent steps imply a progressive reliance on variable renewable energy sources to generate electricity, together with a larger degree of electrification of the industrial, commercial, transportation and domestic sectors [1].

The energy transition is particularly pressing on islands, whose energy supply typically depends on imported fossil fuels and submarine connections to mainland or nearby islands, which increases generation costs [2]. Given their typically small size and limited inter-connectivity, islands have less robust power grids compared to the mainland. In turn, they are prone to more frequent failures. Moreover, many islands rely on tourism as their main economic activity, thus they are subject to seasonal changes of population and large demand variations. Therefore, replacing conventional with renewable generation in these territories requires specific analysis of their operational challenges [3]. In the case of the Balearic Islands, the energy transition has led to the partial close down of its coal fired power plant. We base our analysis on the grid frequency statistics, as has been considered for several locations [4]. Here, we focus on the effect that the replacement of coal has had on this statistics.

The paper is structured as follows. Section II introduces the Balearic grid as our case study. Section III presents the two data sets which will be used. Sections IV and V analyze frequency fluctuations in the absence and presence of coal generation, respectively. Section VI discusses the presence of threshold-like frequency control. Finally, Section VII, summarises the concluding remarks of our study.

II. The Balearic grid

The Balearic Islands are a Spanish archipelago located in the Mediterranean Sea, near the eastern coast of the Iberian Peninsula. Their high-voltage power grid can be mapped down to substation level as a network of 61 nodes and 88 links distributed across its four largest islands, i.e., Mallorca, Menorca, Ibiza, and Formentera [5]. This includes 6 conventional power plants, which we summarize in Table I, and the 3 Alternating Current (AC) submarine interconnections of Menorca-Mallorca, Mallorca-Ibiza, and Ibiza-Formentera. Moreover, Mallorca has a High Voltage Direct Current (HVDC) submarine connection to mainland that provides around 30% of the total demand [6].

The AC connections among the different islands ensure the synchronous operation of the Balearic grid. Since the line with mainland is DC, the Balearic grid operates asynchronously with respect to European continental grid.

Although islands only account for a small fraction of global greenhouse gas emissions, they are one of the most vulnerable territories to the effects of climate change. The Balearic Islands are no exception. For this reason, in 2019, the electric utility company Endesa, the Balearic and the Spanish Government reached an agreement to close down 2 out of the 4 coal generating units of Es Murterar [7], the most polluting power station in Mallorca.

Besides the close down, they also limited the amount of operation time of the two remaining units to 1500 hours

TABLE I. INSTALLED POWER AT EACH CONVENTIONAL POWER PLANT OF THE BALEARIC GRID BEFORE THE PARTIAL CLOSE DOWN OF ES MURTERAR [8]. CCGT STANDS FOR COMBINED CYCLE GAS TURBINE.

Power plant	Generation type	Installed capacity (MW)
Mahón (Menorca)	gas turbine + ancillary	171.7 32.7
	diesel engine (ancillary)	40.8
Es Murterar (Mallorca)	coal	241.2
	gas turbine	65.4
Son Reus (Mallorca)	CCGT	394
	gas turbine	134.8
Cas Tresorer (Mallorca)	CCGT	429
Ibiza	gas turbine + ancillary	119 68
	diesel engine + ancillary	69.6 29
Formentera	gas turbine (ancillary)	11.5

per year until 2021. After that, the number of hours will be reduced to 500 per year until the complete close down of the power plant, which will coincide with the activation of a new connection to mainland. These measures were taken in order to decrease emissions, as a step in the decarbonization agenda [8]. Nowadays, the main technology types in the archilepago are combined cycle, gas turbines, and diesel engines.

III. DATA

Frequency data measured every second is obtained from the open database [9] [4]. The database includes measurements from October 2019 until December 2020, except the months of August and October 2020. The data was taken at a single location in the island of Mallorca, and we assume that the grid frequency is the same in the other islands.

We also make use of data publicly available on the web site of Red Eléctrica de España (REE) [10], the Spanish grid operator, who is responsible for maintaining the demand-supply balance under specific power quality conditions. In particular, for the case of the Balearic Islands, they provide the overall demand and generated power averaged over 10 minutes, as well as the power arriving from mainland Spain through the HVDC line. Generation is disaggregated by power plant technology.

IV. ANALYSIS OF FREQUENCY FLUCTUATIONS

Frequency fluctuations illustrate supply-demand unbalances, which are a result of the unpredictable load changes. According to Spanish legislation [11], the reference frequency for the power grid is 50 Hz and the statutory operational limits are between 49.85 and 50.15 Hz.

Since we have access to both frequency and power data, the first step in our study is to simply compare these two data sets. In Figure 1, we show the comparison for one day in January 2020, when there was no coal generation. In the upper panel, we have the 10-minute power data for demand and generation disaggregated by technology, and in the lower panel, we have the 1-second frequency data.

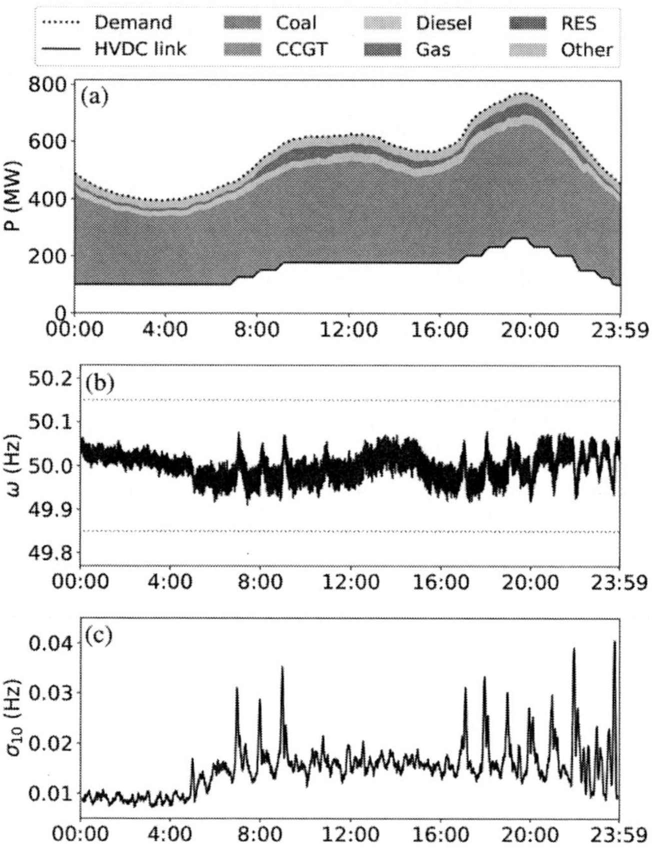

Figure 1. Time evolution of the demand and generation (a) and frequency fluctuations (b) on January 26, 2020, a day in which there was no coal generation. On panel (a), generation is disaggregated by power plant technology, including the HVDC connection to mainland. On panel (b), the dotted lines indicate the statutory operational limits, i.e., (50.00 ± 0.15) Hz. Panel (c) shows the moving standard deviation of the frequency calculated using a 10 minute sliding window.

The first point that we notice is the overall behavior of the frequency, which follows the daily pattern of the demand. During the first hours of the day, the frequency is above 50 Hz indicating an excess of generation. Since the demand is decreasing, so is the generation. However, the generation runs a bit behind because it is a response to the changes in the demand, hence the excess.

Early in the morning, the decrease rate of the demand slows down, which brings the frequency near its nominal value. In other words, the generation closely matches the demand. However, at some point, the demand starts increasing, which causes a lack of generation and makes the frequency drop below 50 Hz. Then, we could follow the same reasoning throughout the rest of the day to see how the slow changes in power are linked to the grid frequency.

Nonetheless, there are also fast power variations, which are naturally responsible for the fast frequency fluctuations. We are referring to the stochastic changes in the demand caused by consumers. Although these changes are not recorded in the 10-minute power data, they can be seen in the frequency data by looking at the thickness of the curve. This is further evidenced in panel (c) which shows the frequency volatility measured by the moving standard deviation σ_{10} evaluated using a sliding time window of duration 10 minutes. As expected, we notice the difference between daytime and nighttime, when there is less frequency volatility because consumers are asleep.

Besides the random fluctuations of the demand, there can also be large deterministic events. These can be both demand or generation power changes induced by a unique and sometimes anomalous cause. In Figure 1, we can see that this is the case of the step-like changes in the power provided by the HVDC connection between Mallorca and mainland Spain, which cause large frequency changes visible in panel (b), also reflected as large peaks in the moving standard deviation σ_{10} (panel (c)). In fact, most of the large frequency changes in that day can be identified with changes in this power line. The size of these large frequency shifts depends, of course, on the power imbalance, but it also depends on the amount of control available in the power plants operating at that particular moment. It should be noted that during low load hours the system is more susceptible due to the decrease in conventional generation, which affects the inertial response. Therefore, a HVDC power change in the morning can have a different impact on the system than if the same variation happened at night.

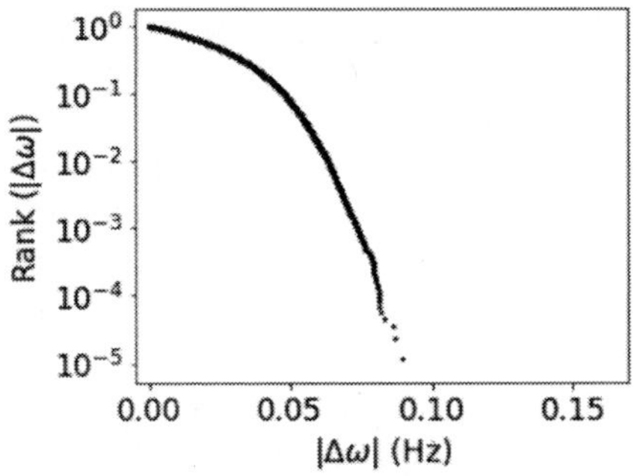

Figure 2. Rank distribution of the absolute value of the frequency fluctuations on January 26, 2020, when there was no coal generation and the energy mix was dominated by CCGT.

In order to characterize the frequency fluctuations, we can use the rank size distribution. We evaluate $\Delta\omega_k \equiv \omega_k - \omega_R$, where ω_R is the reference frequency (50 Hz), reorder the set of values $|\Delta\omega_k|$ from the smallest to the largest value, and finally estimate the complementary cumulative distribution of the deviations as $R(|\Delta\omega_i|) = 1-(i-1)/(M-1)$, where M is the number of data points. $R(|\Delta\omega|)$ measures the probability to have a frequency fluctuation of size larger than $|\Delta\omega|$. In Figure 2, we plot the result. We see that frequency variations from the nominal value stay below 0.1 Hz, which we could already see in Figure 1b. Moreover, the shape of the curve shows a smooth decay in the probability of having large fluctuations. In other words, the frequency tends to stay close to its nominal value, and large deviations are highly unlikely.

V. Coal generation and frequency fluctuations

As we indicated in Section II, the year 2019 marked the end of coal as the main source of power generation in the Balearic Islands. In Figure 3, we plot the daily average power generated from coal (panel a) and from CCGT (panel b) in 2019 and 2020.

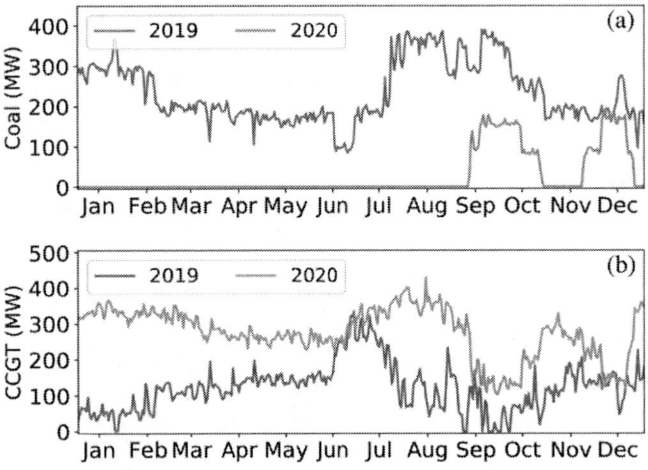

Figure 3. (a) Coal generation daily average in 2019 (gray) and 2020 (red). (b) CCGT generation daily average in 2019 (gray) and 2020 (blue).

To better appreciate the changes in the energy mix disregarding the seasonal variations in demand, we plot in Figure 4 the percentage of generation covered by the different generation technologies in 2019 (panel a) and 2020 (panel b). We can see that coal generation has been replaced by natural gas (combined cycle), which is a less polluting fossil fuel. However, during certain periods of 2020, coal was still used for electricity generation. In fact, it represents a very substantial part of the energy mix on these periods.

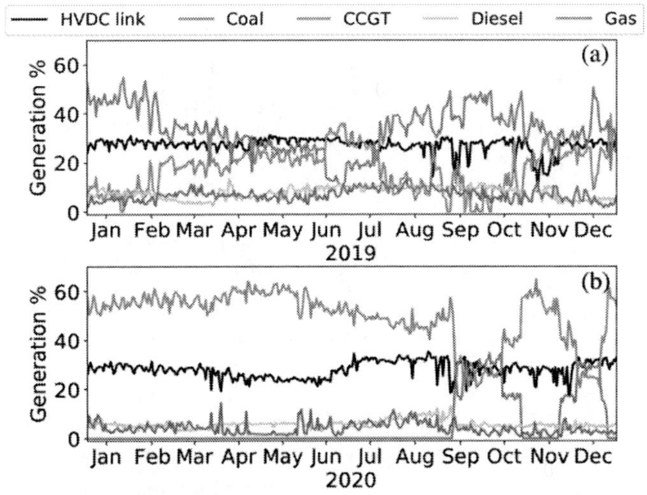

Figure 4. Percentage of the generation covered by different power plant technologies in (a) 2019 and (b) 2020.

In Figure 5, we plot the demand and generation disaggregated by power plant technology (panel a), the grid frequency (panel b), and the frequency volatility (panel c) for a typical day in 2019, when there were no restrictions to coal generation. Looking at panel (b), we observe large frequency deviations, specially compared to those in Figure 1b. In fact, it is clear that the frequency reaches the statutory limits of ± 0.15 Hz on several occasions. The frequency volatility, as indicated by the standard deviation σ_{10}, is also much larger when coal generation dominates the energy mix, as shown in panel (c) (compare this panel with that of Figure 1). Altogether, this is an indication that the overall control capacity of the Balearic grid is significantly smaller than the case shown in Figure 1. We believe that this is simply because combined cycle power plants have a faster and more powerful control response.

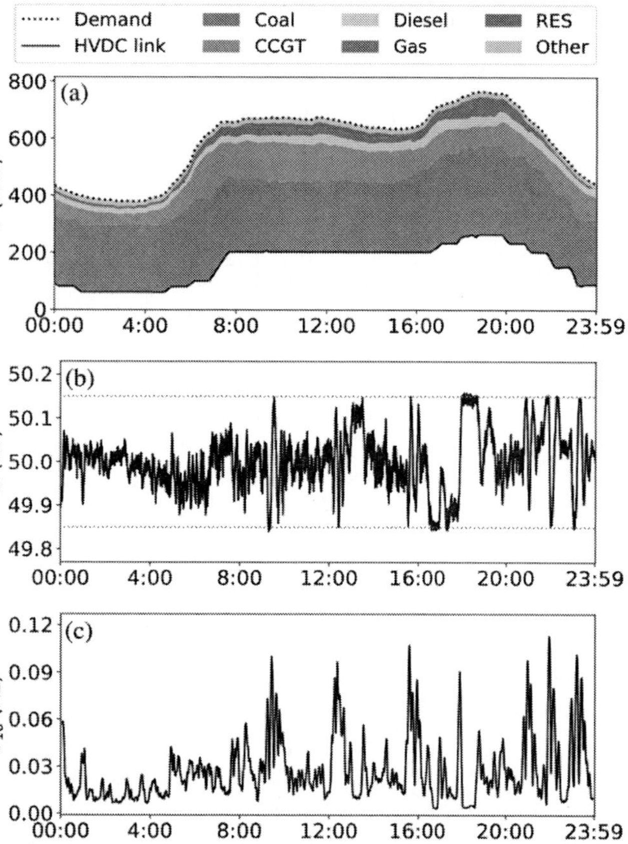

Figure 5. Time evolution of the demand and generation (a) and frequency fluctuations (b) on December 18, 2019, a day in which coal was the main source in the energy mix. On panel (a), generation is disaggregated by power plant technology, including the HVDC connection to mainland. On panel (b), the dotted lines indicate the statutory operational limits, i.e., (50.00 ± 0.15) Hz. Panel (c) shows the frequency volatility σ_{10}.

We also computed the rank distribution of frequency fluctuations for this case, which we show in Figure 6. Comparing it with Figure 2, we can confirm the difference in terms of fluctuation sizes, but also in the shape of the distribution. The smooth parabolic decay that we saw in Figure 2 is not present

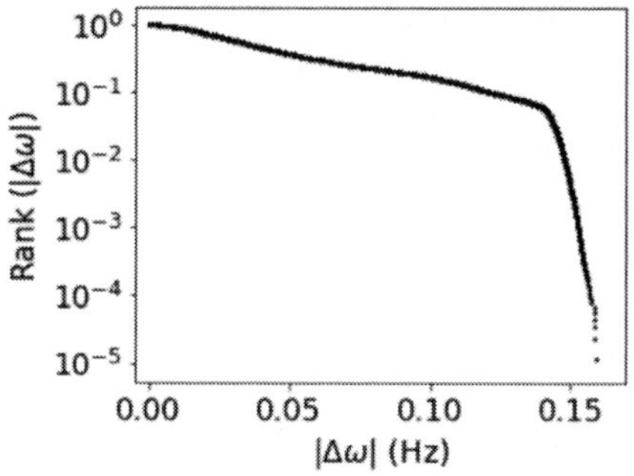

Figure 6. Rank distribution of the absolute value of the frequency fluctuations on December 18, 2019, when coal was the main energy source.

in Figure 6. Instead, we see a much slower linear decay of frequency deviations up to 0.15 Hz, which is followed by a very steep drop. This indicates that the probability to have frequency fluctuations up to 0.15 Hz is significantly larger than in the case considered in Figure 2. Nevertheless, the probability to have frequency fluctuations beyond the statutory limits ± 0.15 Hz is very small.

VI. Discussion on Threshold-like Frequency Control

The behavior of the frequency fluctuations displayed in Figure 5b and Figure 6 is very peculiar. The sharp cut of the frequency deviations at the statutory limit of 0.15 Hz is not observed in the analysis of other power grids [4]. Besides the cut of the frequency deviations at ± 0.15 Hz, we can also see that when the frequency reaches that value, it may remain clamped at that value for a relatively long period of time (tens of minutes or even more than one hour). For instance, this is what happens around 6 PM in Figure 5, when the frequency is kept at its upper limit for 45 minutes.

The typical control mechanisms present in conventional power plants tend to restore the frequency back to its nominal value and act proportionally to the frequency deviation or depending on a smooth function of the frequency deviation. The existence of a threshold-like value beyond which the damping of the fluctuations is much stronger is not the natural response of these mechanisms. For these reasons, we conclude that there must be an additional control which is activated when the frequency deviation reaches the statutory limits ± 0.15 Hz.

The effect of this threshold-like frequency control is more noticeable in periods of coal generation, when there are larger frequency fluctuations. However, we have found that it can also be seen when combined cycle is the main generating technology, although in this case it happens seldomly. This is illustrated in Figure 7 for January 30, 2020. Although there was no coal generation, during the first hours of the day the frequency reached 50.15 Hz and it stayed around that value for 2 hours. As a consequence, in this day the rank distribution of the frequency fluctuations has a sharp drop at $|\Delta\omega| = 0.15$ Hz, as shown in Figure 8.

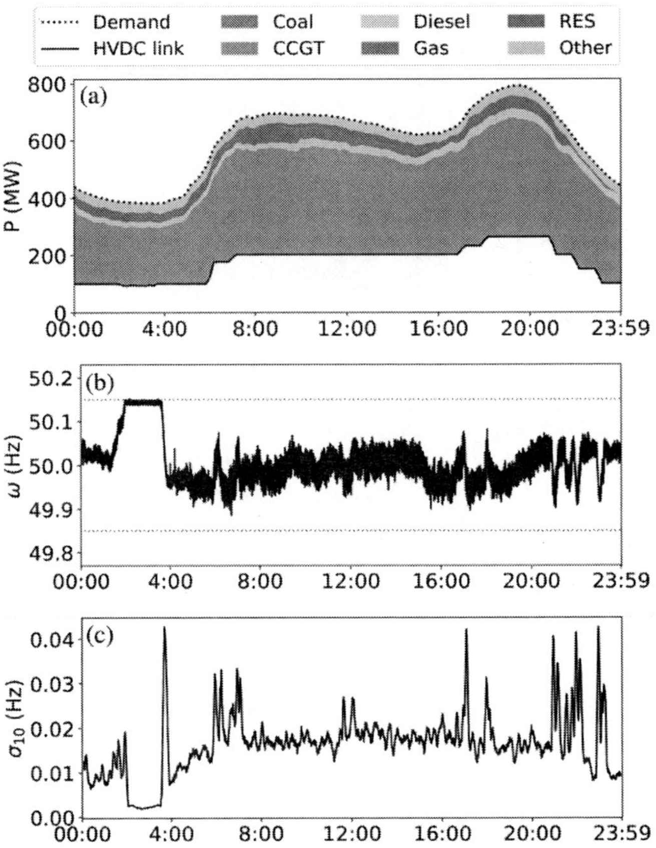

Figure 7. Time evolution of the demand and generation (a) and frequency fluctuations (b) on January 30, 2020. On panel (a), generation is disaggregated by power plant technology, including the HVDC connection to mainland. On panel (b), the dotted lines indicate the statutory operational limits, i.e., (50.00 ± 0.15) Hz. Panel (c) shows the frequency volatility σ_{10}.

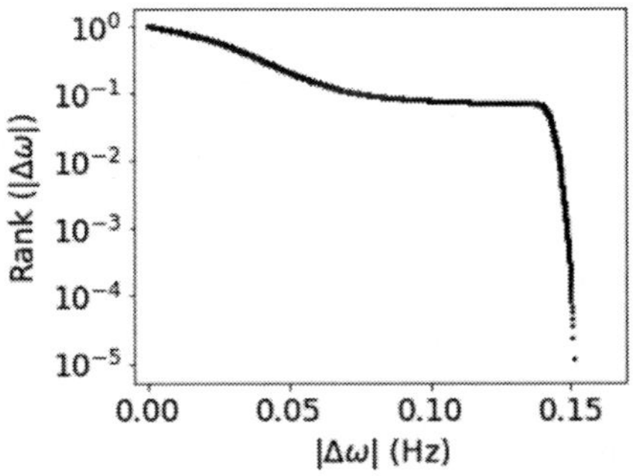

Figure 8. Rank distribution of the absolute value of the frequency fluctuations on January 30, 2020, when the threshold-like frequency control was activated despite not having coal generation.

VII. CONCLUSIONS

We have analyzed the frequency fluctuations recorded in the Balearic power grid. We have seen that the partial close down of its coal fired power plant in 2019, being replaced by CCGT, has lead to significant reduction of the frequency fluctuations. Nowadays, CCGT is the main generating technology during large part of the year. However, coal can still be used for electricity generation. On the periods of time in which this is the case, the frequency is more volatile and has larger fluctuations.

In Figure 9, we compare the rank size distribution of frequency variations for a three-months period in which coal was the main energy source (red) with a similar period where it was replaced by CCGT (blue). There is a clear difference between the two scenarios. When there is coal generation, we observe a similar shape to that in Figure 6, with a roughly power-law decay up to 0.15 Hz followed by a sharp decrease all the way down to cumulative probabilities of the order of 10^{-6}. This sharp decay is associated to the activation of threshold-like frequency control, which happens quite frequently for periods of time in which coal is the main component of the energy mix. The sharp decay is followed by another power-law decay for fluctuations larger than 0.2 Hz, which was not visible on Figure 6 since it is associated to very rare events not present on the particular day of the figure.

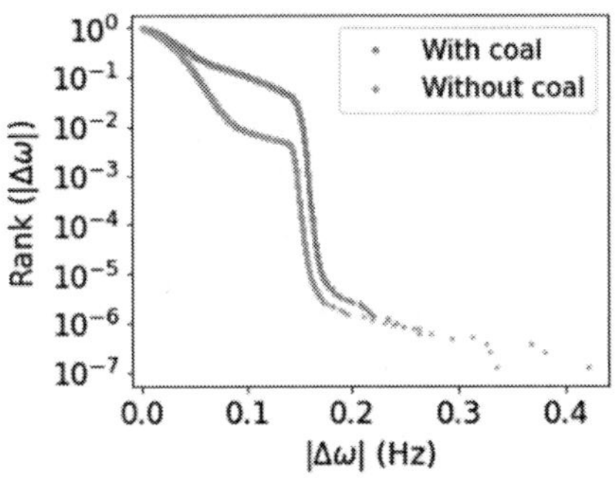

Figure 9. Rank distribution of frequency deviations measured from October to December 2019 (red), and from January to March 2020 (blue).

When there is no coal generation, the shape of the distribution decays faster than power law for $|\Delta\omega| < 0.10$ Hz. This fast decay is followed by a practically horizontal plateau up to 0.15 Hz which indicates that there are very few fluctuations of that size. This is what we saw in Figure 8, and it has to do with the fact that the frequency stays within $(49.9, 50.1)$ Hz most of the time. However, larger frequency variations can occur as illustrated in Figure 7b, where the frequency goes up to 50.15 Hz and, after some time, it jumps back down to a point closer to its nominal value. The point is that it spends very little time in the range $(50.10, 50.15)$ Hz, which is why the size distribution is flat around those values. Once the frequency fluctuations reach 0.15, the threshold-like mechanism is activated, albeit this happens seldomly when there is no coal generation and the energy mix is dominated by CCGT, leading to a sharp decay in the rank distribution. Finally, the power-law tail associated to fluctuations larger than 0.2 Hz is still present but its probability is lower than in the case of coal-dominated generation mix.

To further illustrate the differences in the frequency statistics when coal or CCGT are the main generation source, in Figure 10, we plot the probability density function for the daily time series that we have analyzed in Figures 1, 5, and 7. TABLE

II shows the mean, variance, skewness, and kurtosis of these data sets. The data set in Figure 1 has the mean closest to 50 Hz, and the smallest variance, indicating that frequency fluctuations are smaller than for the other cases. This data set has also the most symmetric distribution (smallest skewness). Nonetheless, as we can see in Figure 10 and with a skewness of 0.09, the data in Figure 5 can also be considered symmetric. In contrast, the data set in Figure 7 is highly skewed to the right. This is due to the fact that the frequency fluctuates around its reference value for most of the time, except during the two hours where it stays in the limit 50.15 Hz. For the kurtosis, we have to compare it to that of the normal distribution, i.e., 3. We see that Figure 1 is platykurtic (less than 3) and Figure 7 is leptokurtic (greater than 3). Therefore, in the former, the probability of large events is smaller than in the Gaussian distribution, while it is larger in the latter.

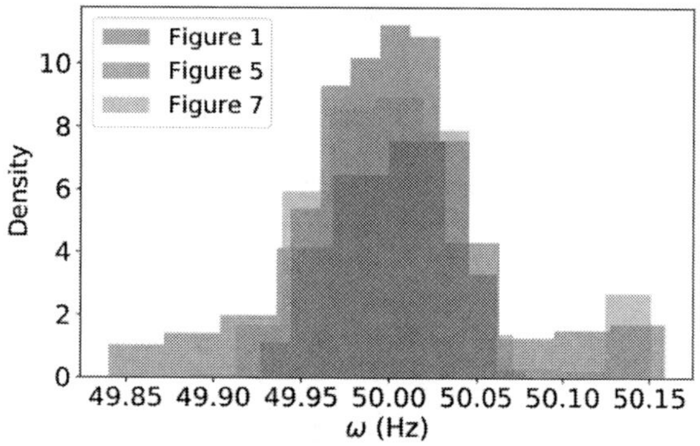

Figure 10. Probability density function of the daily frequency time series shown in Figures 1, 5, and 7. The histogram is normalized so that the total area is 1, as it corresponds for a probability density function.

TABLE II. MEAN, VARIANCE, SKEWNESS, AND KURTOSIS OF THE FREQUENCY TIME SERIES SHOWN IN FIGURES 1, 5, AND 7.

Measure	Figure 1 2020-01-30	Figure 5 2019-12-18	Figure 7 2020-01-30
Mean	49.999	50.002	50.006
Variance	0.03	0.06	0.05
Skewness	−0.03	0.09	1.17
Kurtosis	2.19	3.04	4.50

Moreover, we plot in Figure 11 the probability density function of the frequency fluctuations obtained from sampling the frequency every second during a three-months measurement period. The probability distribution for the coal-dominated period (red) shows a much longer tail than the one for the CCGT-dominated period. As a consequence, while frequency deviations between 0.10 and 0.15 Hz are quite probable for the former case, they are very rare for the last one.

Finally, we have also seen that there is clear evidence of a threshold-like frequency control in addition to the typical control mechanisms operating in conventional power plants. This additional control is activated when the frequency deviation reaches the statutory limits ± 0.15 Hz, strongly damping deviations beyond this threshold. Its effect can be seen in the frequency time series (Figure 5 and Figure 7), and also as a steep cut in the rank size distribution (Figure 9).

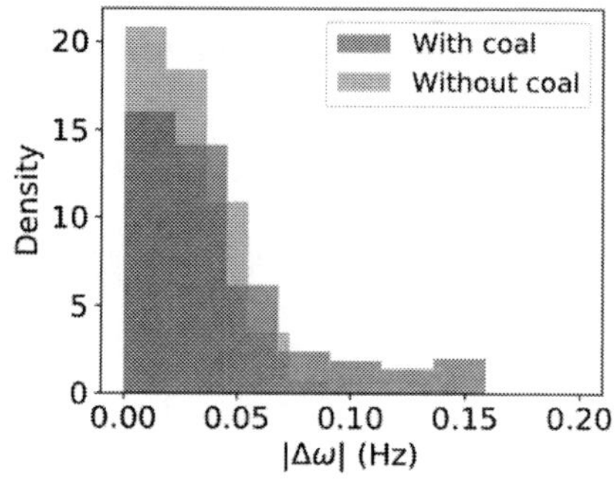

Figure 11. Probability density function of frequency deviations measured from October to December 2019 (red), and from January to March 2020 (blue). The histogram is normalized so that the total area is 1, as it corresponds for a probability density function.

ACKNOWLEDGEMENTS

We acknowledge financial support from Ministerio de Ciencia e Innovación (Spain), Agencia Estatal de Investigación (AEI, Spain), and Fondo Europeo de Desarrollo Regional (FEDER, EU) under grant PACSS (RTI2018-093732-B-C22) and the Maria de Maeztu program for Units of Excellence in R&D (MDM-2017-0711) and from European Comission integrated action Virtual Power Plants for Islands (VPP4ISLANDS).

REFERENCES

[1] International Energy Agency, "World Energy Outlook 2020". Available: https://www.iea.org/reports/world-energy-outlook-2020. [retrieved: May, 2021]

[2] D. Weisser, "On the economics of electricity consumption in small island developing states: a role for renewable energy technologies?," *Energy Policy*, vol. 32, no.1, pp. 127–140, 2004.

[3] O. Erdinc, N.G. Paterakis, and J.P.S. Catalão, "Overview of insular power systems under increasing penetration of renewable energy sources: Opportunities and challenges," *Renew. Sustain. Energy Rev.*, vol. 52, pp. 333–346, 2015.

[4] L.R. Gorjão et al., "Open database analysis of scaling and spatio-temporal properties of power grid frequencies," *Nat. Commun.*, vol. 11, no. 1, pp. 1–11, 2020.

[5] Red Eléctrica de España, "Singular features of the Balearic system". Available: https://www.ree.es/en/activities/balearic-islands-electricity-system/singular-features-of-the-system. [retrieved: May, 2021]

[6] Red Eléctrica de España, "Rómulo Project". Available: https://www.ree.es/en/activities/unique-projects/romulo-project. [retrieved: May, 2021]

[7] Majorca Daily Bulletin, "Partial closure of Alcudia power station formally authorised". Available: https://www.majorcadailybulletin.com/news/local/2019/03/29/55007/partial-closure-alcudia-power-station-formally-authorised.html. [retrieved: May, 2021]

[8] Endesa, "Non-peninsular territories 100% decarbonised by 2040". Available: https://www.endesa.com/en/press/press-room/news/energy-transition/transition-non-peninsular-territories-2040. [retrieved: May, 2021]

[9] R. Jumar, H. Maass, B. Schäfer, L.R. Gorjão, and V. Hagenmeyer, "Power grid frequency data base". Available: https://osf.io/by5hu/. [retrieved: May, 2021]

[10] Red Eléctrica de España. Available: https://demanda.ree.es/visiona/home. [retrieved: May, 2021]

[11] Resolution 20053 of June 30, 1998 of the Spanish Secretaría de Estado de Energía y Recursos Minerales, published in *Boletín Oficial del Estado* on August 8, 1998. https://www.boe.es/eli/es/res/2018/02/01/(3). [retrieved: May, 2021]

The Risk of Cascading Failures in Electrical Grids Triggered by Extreme Weather Events

Julian M. Stürmer
(1) Complexity Science department,
Potsdam Institute for
Climate Impact Research
Potsdam, Germany
(2) Institut für Theoretische Physik, TU Berlin
Berlin, Germany
email: Julian.Stuermer@pik-potsdam.de

Anton Plietzsch
(1) Complexity Science department,
Potsdam Institute for
Climate Impact Research
Potsdam, Germany
(2) Humboldt-Universität zu Berlin
Berlin, Germany
email: plietzsch@pik-potsdam.de

Mehrnaz Anvari
Complexity Science department,
Potsdam Institute for
Climate Impact Research
Potsdam, Germany
email: anvari@pik-potsdam.de

***Abstract*—One of the serious threats related to climate change is an increase in the number and severity of extreme weather events. A prominent example are hurricanes, which result from rising coastal temperatures. Such extreme weather events can cause extensive damages in infrastructure systems and, potentially, destroy components in electricity transmission networks, which in turn can lead to major blackouts. In our recent work, we consider a quasi-static model to study the risk of hurricane-induced cascading failures in power systems of the U.S. East Coast using historical wind field data sets. For this purpose, we model the destruction of overhead transmission lines during hurricanes, where each failing line causes a rerouting of power flow in the system. New power flows can overload additional lines, which are then automatically deactivated and thereby cause another rerouting of power flow and so on. Ultimately, a cascade of failures can unfold that can black out large parts of the power system.**

***Keywords*—weather extreme events; power flow model; cascading failure.**

I. Introduction

Recent climate studies demonstrate that human-induced climate change not only causes a rapid increase in global temperature but also leads to more frequent and severe extreme weather events, such as floods, heavy rainfall, winter storms and hurricanes [1]. These events threaten the stability of social facilities and services, as well as social network infrastructures, such as public health care, transportation, telecommunication, and electrical grids. Due to the extensive economic and social consequences that accompany disruptions in these systems, fostering their resilience and thereby mitigating the impact of extreme weather events represents an important challenge for governments and societies.

It is worth mentioning that, due to the intertwined nature of social infrastructure systems, a failure in one system can easily spread into other systems. This especially applies to failures in electrical grids, since they can lead to major blackouts that impair the access to food, transportation, medical treatment and so on. In recent years, reports demonstrated that the increase in extreme weather events puts electrical grid components at greater risk of failure in several parts of the world. As an example, the authors in [2] discuss how the probability of damages in the British transmission network increases with more frequent winter storms. In addition, [3] shows how transmission tower damages in Australia can be attributed to localized downbursts, that become more often.

Furthermore, recent data recorded in the Atlantic Ocean [4], [5] show an increase in coastal temperatures that promote more frequent and intense hurricanes in coastal states of the U.S. like Texas and Louisiana. For instance, hurricane Laura led to major outages in Louisiana and other states in the days following the August 27th, 2020 [6]. Since the restoration of destroyed components in electrical grids can be very costly for power transmission authorities (typically ranging from millions to billions of euros), identifying vulnerable components and improving the resilience represent a promising way of avoiding high costs and extensive outages.

In our ongoing research, we apply a quasi-static model to investigate the risk of hurricane-induced blackouts in a synthetic electrical grid for Texas. In Section II, we discuss how power flows can be analyzed in the electrical grid. Next, we establish a probabilistic approach in Section III to model wind-induced failures of overhead transmission lines occurring in the course of a traversing hurricane. Our work incorporates the analysis of power flows that change after each failure and the simulation of cascading failures by deactivating lines and transformers as soon as they become overloaded. Preliminary results of blackouts will be presented in Section IV.

II. Electrical Grid Modeling

To assess the impact of extreme weather events in realistic scenarios, an elementary question arising is how and in what detail the electrical grid should be modeled. Regarding to the power system research, an electrical grid is a network with N nodes, representing generators and loads, and E edges, representing transmission lines and transformers. We will also refer to nodes as buses and to edges as branches throughout this paper.

In the context of cascading failures, both steady-state models, as well as dynamic models have been used for calculating the power flow on branches and it is known that the final outcome of cascades can vary between different models [7]. It is therefore not certain how cascading failures should be modeled in general in order to obtain the most realistic and plausible results. Furthermore, the applicability of different power flow models also depends on the availability of grid parameters, such as inertia and damping coefficients in the dynamic swing equation or reactive power injections in the static AC power flow model. Data sets that provide these parameters together with a realistic power grid topology are generally rare because of confidentiality.

The application presented throughout this paper uses a sophisticated synthetic electrical grid data set generated by Birchfield et al. on the footprint of Texas [8]. The data set is publicly available under the name "ACTIVSg2000" in a test case repository hosted by the Texas A&M University [9]. It encompasses 2000 buses with geographic locations and an extensive set of parameters, some of which will be introduced in Section II-A.

The following sections will introduce the AC power flow model and its DC approximation, both of which allow to calculate power flows in the regarded electrical grid of Texas. Using these models to determine the steady states of the power grid enables a quasi-static description of cascading failures that will be discussed in Section IV.

A. AC Power Flow

The AC power flow equations representing a set of $2N$ nonlinear algebraic equations are solved to obtain the voltage magnitudes $|V_i|$ and voltage angles $|\theta_i|$ of all buses $i \in \{1,\ldots,N\}$. These voltage variables characterize the steady state of an AC power grid and enable the calculation of all power flows. The AC power flow equations are given by

$$P_i = \sum_{j=1}^{N} |V_i||V_j|[G_{ij}\cos(\theta_i - \theta_j) + B_{ij}\sin(\theta_i - \theta_j)]\,, \quad (1)$$

$$Q_i = \sum_{j=1}^{N} |V_i||V_j|[G_{ij}\sin(\theta_i - \theta_j) - B_{ij}\cos(\theta_i - \theta_j)]\,, \quad (2)$$

where $P_i \in \mathbb{R}$ and $Q_i \in \mathbb{R}$ represent the active and reactive power injections at each bus. The coefficients $G_{ij} \in \mathbb{R}$ and $B_{ij} \in \mathbb{R}$ are, respectively, the real and imaginary part of the nodal admittance matrix $Y_{ij} = G_{ij} + \mathrm{i}B_{ij} \in \mathbb{C}$ and incorporate branch conductances and susceptances as well as additional shunt contributions. All of these parameters are provided in the "ACTIVSg2000" data set and their values are generally given in the per-unit system (p.u.). In order to solve the AC power flow equations (1) and (2), we draw upon the Julia package PowerModels.jl [10].

B. DC Approximation

The DC approximation represents a linearized version of the AC model introduced in Section II-A. Following three key assumptions have to be made for DC power flow: (i) Neglect ohmic losses resulting from resistances, i.e. $G_{ij} = 0\,\mathrm{p.u.}$ for all branches in the network; (ii) Fix the voltage magnitudes to $|V_i| = 1\,\mathrm{p.u.}$ for all buses $i \in \{1,\ldots,N\}$. This reduces the number of unknown state variables by N and one can therefore drop the N reactive power equations shown in (2); (iii) Consider small voltage angle differences $|\theta_i - \theta_j|$ along all branches. This allows us to linearize the sine function in (1) and, ultimately, leads to the N equations of DC power flow.

$$P_i = \sum_{j=1}^{N} B_{ij}(\theta_i - \theta_j)\,. \quad (3)$$

These equations are also solved for the considered Texas grid using PowerModels.jl. Due to the strong approximations made in the DC model, the resulting power flows tend to differ from the AC solution. A comparison of both solutions is shown in Figure 1, where we plot the loading of branches in the DC model versus their corresponding loading in the AC model. As the ACTIVSg2000 test case for Texas only contains high voltage levels of $115\,\mathrm{kV}$ and higher, ohmic losses can be insignificant and, therefore, assumption (i) should not lead to a major inaccuracy of the DC model. Nevertheless, Figure 1 shows that various branches transport considerably more power in the AC model. This can only stem from the fact that reactive power flows, which also contribute to the loading in the AC model, were neglected in the DC model according to assumption (ii). In the context of cascading failures driven by overloaded branches, this can lead to substantially different cascades in both models. Last, assumption (iii) can result in an overestimation of flows in the DC model, if voltage angle differences are not small along all branches. Figure 1, however, shows that this overestimation is not as drastic as the negligence of reactive power flows.

Motivated by the discussed discrepancy between the initial AC and DC solution, we are investigating how this may affect cascading failures in both models in ongoing work. It should be stressed that the DC model generally provides better solvability and can be solved much faster than the AC model, which makes it a popular choice for computing cascading failures. Section IV presents preliminary results using the DC model.

III. Wind-Induced Damage

To demonstrate the impact of extreme weather events on electrical grids, we here discuss the modeling of wind-induced damages. In principle, as wind speeds exceed the design wind speeds of structures in an electrical grid, damages can be expected to occur in the course of an extreme event. However, it is never certain in advance which components will indeed fail. Hence, probabilistic approaches modeling the fragility of structures with regard to weather-induced failures have been developed throughout the years.

For the hurricane scenarios that we study, we adopt a probabilistic description of the fragility of overhead transmission lines first introduced by Winkler et al. in [11]. The method is based on the standard wind force design equation defined by the American Society of Civil Engineers (ASCE) in [12] that allows to calculate the wind force F_{wind} acting on transmission line segments. Line segments represent pairs

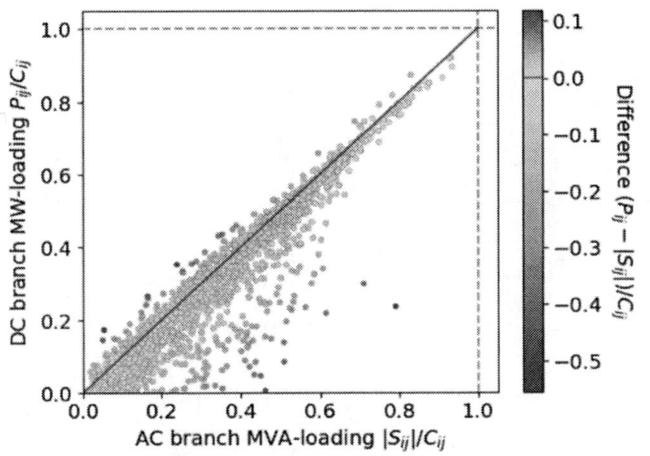

Figure 1. Loading of branches in the DC model versus respective loading in the AC model. Individual data points are colored according to the difference in loading. The diagonal green line indicates the perfect match.

of transmission towers with conductor wires spanned between them. In reality, all overhead transmission lines in the synthetic electrical grid of Texas consist of a large number of line segments. We therefore divide all overhead transmission lines into N_{seg} individual segments using the average distance of $161\,\text{m}$ between transmission towers [13]. Motivated by [11], we proceed by assigning failure probabilities to individual line segments k according to

$$p_k(v,l) = \min\left(\gamma\,\frac{F_{\text{wind},k}(v,l)}{F_{\text{brk},k}},\,1\right)\,, \tag{4}$$

where v is the local 3-second gust wind speed assumed to act perpendicular to the line direction, l is the obtained wire span length of roughly $161\,\text{m}$ and $F_{\text{brk},k}$ is the maximum perpendicular force that the wire can endure, which is chosen according to the grid data set. The coefficient γ is a scaling parameter proposed in [11], that can be used to match the average number of wind-induced line failures to historical data. The impact of the scaling parameter will be discussed in the last paragraph of this section. According to the wind force F_{wind} defined in [12], the failure probabilities in (4) increase with the square of the wind speed $p_k \propto v^2$.

Given the wind speed data, v, for a specific hurricane, we can generate uniform random numbers in $[0,1]$ for each line segment and time step and compare them to the respective probabilities (4). Once a random number falls below the failure probability, we remove the corresponding transmission line containing the destroyed segment. Ultimately, each scenario of a hurricane traversing the grid represents a chronological sequence of wind-induced line failures used to compute cascading failures that may be triggered. As our model is probabilistic, we can perform Monte Carlo simulations to assess the expected damage of the grid for a specific wind data set.

Ultimately, Figure 2 shows that the mean number of destroyed overhead transmission lines increases monotonically with the scaling parameter using the wind field of hurricane Ike. Hurricane Ike will be discussed in more detail in Section IV. As seen in the inset of Figure 2 for an exemplary value of $\gamma = 0.005$, the corresponding probability densities of the number of destroyed lines are narrow and centered around the respective mean. That this observation holds for other values of the scaling parameter is indicated by the rather small error bars. Figure 2 suggests that it should indeed be possible to fit the mean number of wind-induced line failures to historical data. However, finding reliable sources that summarize such data turned out to be a difficult task, since grid operators do not report on structural damages in an uniform format.

IV. Cascading Failures

In this section, we explore cascading failures that unfold during specific hurricane scenarios described in section III. For this purpose, the power flow models introduced in Section II are used to recalculate power flows every time transmission lines are destroyed by the traversing hurricane. We also deactivate overloaded branches that may arise in new power flow solutions and repeat this process until no more overloaded branches persist. Afterwards, our simulations advance to the next wind-induced line failure and so on. This approach can be justified by the fact that cascading failures driven by overloads typically happen on the time scale of seconds while significant changes in the wind speeds typically occur in the course of minutes to hours.

In order to prevent infeasible power flow problems during our simulations, we implement additional control loops that enclose the power flow calculation. This is especially important in the AC model, whenever reactive power is no longer generated in the direct neighborhood of loads. Since refining these loops is still work in progress, we continue by focusing on the simpler DC model. The latter merely requires a single outer loop restoring active power balance

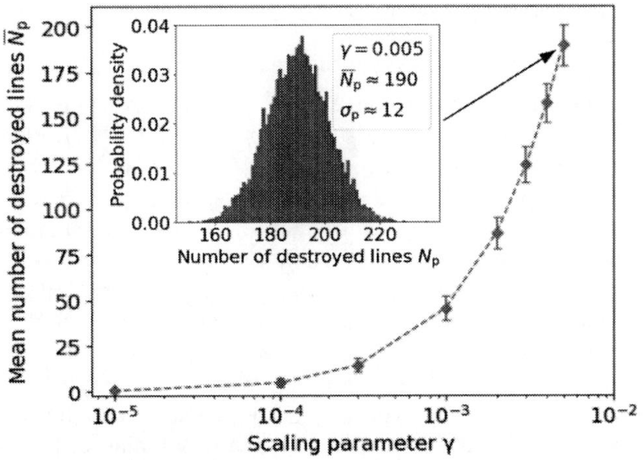

Figure 2. Influence of the scaling parameter introduced in (4) on the mean number of overhead transmission lines destroyed by hurricane Ike. The inset shows the probability density of the number of destroyed lines for an exemplary value of $\gamma = 0.005$.

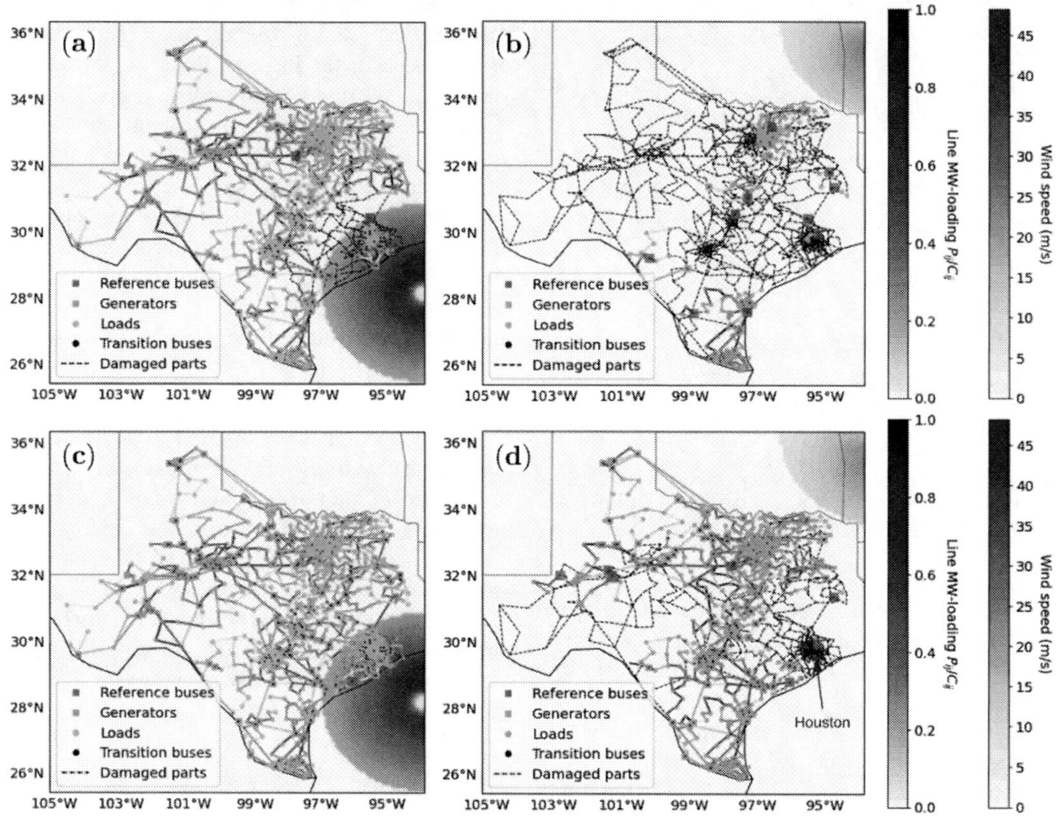

Figure 3. Two different cascading failure scenarios triggered by hurricane Ike. (a) and (b) show two time steps in a worst case scenario. (c) and (d) show the same time steps in a typical scenario. Houston which is extensively damaged in both worst and typical scenario has been annotated in (d).

after generators and/or loads were disconnected from the grid or the system split into separated islands. Here, active power balance means that the generators cover the total demand in the grid. We have therefore established an algorithm that restores balance motivated by primary frequency control in real systems or deactivates connected components, if balance cannot be restored within the capacity of generators.

Figure 3 shows two time steps in two scenarios of hurricane Ike that passed over the eastern part of Texas in 2008. The wind data was calculated using the CLIMADA Python implementation [15] and the IBTrACS archive [16]. Both scenarios were generated using a scaling factor of $\gamma = 0.005$ corresponding to an average number of 189 wind-induced failures. Figure 3(a) and (b) show a worst case scenario that was found by running 10^4 simulations. The cascading failures triggered in this scenario lead to a final loss of $\sim 81\%$ of the initial active power supply in the grid. In contrast, in the scenario shown in Figure 3(c) and (d) $\sim 33\%$ of the initial supply is lost, which seems to be the most probable outcome judging from our 10^4 runs. It is worth mentioning that, even though hurricane Ike merely hits the eastern part of the grid, the majority of scenarios induce failures in the western part of the grid due to the spreading of overloads. Figure 3 shows that the outcome of different scenarios can vary drastically and we are investigating what can trigger worst case scenarios like the one discussed. It is worth to mention that Houston is extensively damaged and faces a blackout in both the worst case scenario as well as the typical scenario, see Figure 3(b) and (d). Actually, this has been reported by references that recap the real damages caused by hurricane Ike in 2008, such as [14]. This proves the ability of our quasi-static model to capture the impact of a hurricane on an electrical grid.

V. Conclusions

In this paper, we demonstrated that overhead transmission lines destroyed by hurricane Ike trigger cascading failures in a synthetic electrical grid for Texas (see Figure 3). As outlined in Section IV, we combine a Monte Carlo-like method modeling wind-induced failures (see Section III) with a DC power flow model to assess the final power outage in different hurricane scenarios.

As an example, we presented a worst case scenario of hurricane Ike in Figure 3, in which more than 80% of the initial power supply is lost, together with a most probable scenario, in which around 30% is lost. In order to prevent the DC power flow problem from becoming infeasible as the electrical grid is destroyed, we applied a control loop enclosing the power flow calculations that restores active power balance motivated by primary frequency control.

Apart from the DC model, we also touched on the more accurate AC model that requires rather sophisticated control strategies to remain feasible during cascading failures. We show in Figure 1 that AC and DC power flows already differ considerably in the stable initial state of the Texas grid. Motivated by this fact we plan to study different hurricane scenarios using the AC model and compare the final outages to corresponding DC simulations. Here, we only considered an electrical grid of Texas and wind data belonging to hurricane Ike as a case study. Nevertheless, our approach can be applied to electrical grids in other geographical regions, such as South Carolina or Louisiana, and to other hurricanes as well. Moreover, in our future work we will investigate the effect of the electrical grid topology and transmission lines capacity in order to minimize the number of consumers affected by power outages during hurricanes.

Acknowledgment

This project has received funding from the CoNDyNet2 project under grant no. 03EF3055F.

References

[1] P. Stott, "How climate change affects extreme weather events," Science,vol. 352, pp. 1517–1518, 2016.

[2] K. Murray, and K. R. Bell, "Wind related faults on the GB transmission network,", 2014 International Conference on Probabilistic Methods Applied to Power Systems (PMAPS). IEEE, pp. 1–6, 2014.

[3] S. E. Oliver, W. W. Moriarty, and J. D. Holmes, "A risk model for design of transmission line systems against thunderstorm downburst winds," Engineering structures, vol. 22, pp. 1173–1179, 2000.

[4] K. E. Trenberth, L. Cheng, P. Jacobs, Y. Zhang, and J. Fasullo, "Hurricane Harvey links to ocean heat content and climate change adaptation," Earth's Future, vol. 6, pp. 730–744, 2018.

[5] Y. K. Lim, S. D. Schubert, R. Kovach, A. M. Molod, and S. Pawson, "The roles of climate change and climate variability in the 2017 Atlantic hurricane season,", Scientific reports, vol. 8, pp. 1–10, 2018.

[6] U.S. Department of Energy, "Combined report: Hurricanes Sally & Laura ," September 2020, https://www.energy.gov/ceser/downloads/combined-report-hurricanes-sally-laura-9252020 [retrieved: May, 2021].

[7] B. Schäfer, D. Witthaut, M. Timme, and V. Latora, "Dynamically induced cascading failures in power grids," Nat. Commun., vol. 9, pp. 1-13, May 2018.

[8] A. B. Birchfield, T. Xu, K. M. Gegner, K. S. Shetye, and T. J. Overbye, "Grid structural characteristics as validation criteria for synthetic networks," IEEE Trans. Power Syst., vol. 32, pp. 3258–3265, July 2017.

[9] Texas A&M University, "Electric grid test case repository," https://electricgrids.engr.tamu.edu/electric-grid-test-cases/ [retrieved: April, 2021].

[10] C. Coffrin, R. Bent, K. Sundar, Y. Ng, and M. Lubin, "Powermodels.jl: An open-source framework for exploring power flow formulations," 2018 Power Systems Computation Conference (PSCC), pp. 1–8, June 2018.

[11] J. Winkler, L. Dueñas-Osorio, R. Stein, and D. Subramanian, "Performance assessment of topologically diverse power systems subjected to hurricane events," Reliab. Eng. Syst. Saf., vol. 95, no. 4, pp. 323–336, April 2010.

[12] American Society of Civil Engineering, "Substation structure design guide," ASCE Manuals and Reports on Engineering Practice No. 113, July 2008.

[13] E. B. Watson and A. H. Etemadi, "Modeling electrical grid resilience under hurricane wind conditions with increased solar and wind power generation," IEEE Trans. Power Syst., vol. 35, pp. 929–937, March 2020.

[14] S. Mukherjee, R. Nateghi, and M. Hastak, "Data on major power outage events in the continental US," Data in brief, vol. 19, pp. 2079–2083, August 2018.

[15] G. Aznar-Siguan and D. N. Bresch, "CLIMADA v1: A global weather and climate risk assessment platform," Geosci. Model Dev., vol. 12, no. 7, pp. 3085–3097, July 2019.

[16] K. R. Knapp, M. C. Kruk, D. H. Levinson, H. J. Diamond, and C. J. Neumann, "The international best track archive for climate stewardship (IBTrACS): Unifying tropical cyclone data," Bull. Am. Meteorol. Soc., vol. 91, no. 3, pp. 363–376, March 2010.

Modelling Power Grids as Pseudoadaptive Networks

Rico Berner
Institut für Theoretische Physik
Institut für Mathematik
Technische Universität Berlin
Berlin, Germany
ORCID 0000-0003-4821-3366

Serhiy Yanchuk
Institut für Mathematik
Technische Universität Berlin
Berlin, Germany
ORCID 0000-0003-1628-9569

Eckehard Schöll
Institut für Theoretische Physik
Technische Universität Berlin
Berlin, Germany
Potsdam Institute for Climate Impact Research
Potsdam, Germany
ORCID 0000-0002-7318-2672

***Abstract*—Power grids, as well as neural networks with synaptic plasticity, describe real-world systems of tremendous importance for our daily life. The investigation of these seemingly unrelated types of dynamical networks has attracted increasing attention over the last decade. In this work, we exploit the recently established relation between these two types of networks to gain insights into the dynamical properties of multifrequency clusters in power grid networks. For this, we consider the model of Kuramoto-Sakaguchi phase oscillators with inertia and describe the emergence of multicluster states. Building on this, we provide a new perspective on solitary states in power grid networks by introducing the concept of pseudo coupling weights.**

***Index Terms*—power grids, solitary states, sychronization, adaptive networks, phase oscillators with inertia**

I. Introduction

Complex networks describe various processes in nature and technology, ranging from physics and neuroscience to engineering and socioeconomic systems. Of particular interest are power systems, as well as micro and macro power grids [1], [2]. It was shown that simple low-dimensional models capture certain aspects of the short-time dynamics of power grids very well [3]–[5]. In particular, the model of phase oscillators with inertia, also known as swing equation, has been widely used in works on synchronization of complex networks [6], [7] and as a paradigm for the dynamics of modern power grids [8]–[22].

Over the last years, studies on models of oscillators with inertia have revealed a plethora of common dynamical scenarios with adaptive network models of coupled oscillators. These scenarios include solitary states [15], [18], [19], [23], frequency clusters [24]–[27], chimera states [28], [29], hysteretic behavior and non-smooth synchronization transitions [30]–[32]. Moreover, hybrid systems with phase dynamics combining inertia with adaptive coupling weights have been investigated, for instance, to account for a changing network topology due to line failures [33], to include voltage dynamics [34] or to study the emergence of collective excitability and bursting [35].

Despite the apparent qualitative similarities of the two types of models, only recently their quantitative relationship has been discovered [36]. In this paper, we show implications of the relation for dynamical power grid models and provide future perspectives for the research on power grid stability and control.

The paper is organized as follows. In Section II, we introduce the Kuramoto-Sakaguchi model with inertia. In the subsequent section III, we briefly review the analytic relation between power grid models and adaptive networks and introduce the concept of the pseudo coupling matrix. In Section IV we show the emergence of a multicluster for oscillators with inertia. Subsequently, in Section V, we show how the concept of pseudo coupling weights can be used to study solitary states in realistic power grid networks. Finally, in Section VI, we summarize the results and give an outlook.

II. Kuramoto-Sakaguchi model with inertia

The model that is considered throughout this paper is given by N coupled phase oscillators with inertia [15]

$$M\ddot{\phi}_i + \gamma\dot{\phi}_i = P_i - \sigma\sum_{j=1}^{N} a_{ij}\sin(\phi_i - \phi_j + \alpha), \qquad (1)$$

with phase $\phi_i(t) \in (0, 2\pi]$ and phase velocity $\dot{\phi}_i(t) = \frac{\mathrm{d}\phi_i(t)}{\mathrm{d}t}$ of the ith node ($i = 1, \dots, N$), corresponding to generators and loads [34]. The $\phi_i(t)$, $\dot{\phi}_i(t)$ are defined relative to a rotation with reference power line frequency ω_G, e.g., 50 Hz for European power grid. The parameter M is the inertia coefficient, γ is the damping constant, σ is the overall coupling strength, and P_i is the power injected or consumed at node i (related to the natural frequency $\omega_i = P_i/\gamma$). The connectivity between the oscillators is described by the entries $a_{ij} \in \{0, 1\}$ of the adjacency matrix A. Further, the coupling function $\sin(\phi + \alpha)$ is parameterized by the phase lag parameter α [37]. The phase lag can be interpreted as part of complex impedance [19].

III. Relating power grid models to adaptive networks

In the following, we present a relation between phase oscillator models with inertia and systems with adaptive coupling weights, and provide an extension for higher order power grid models including voltage dynamics.

A. Pseudo coupling weights: The link between inertia and network adaptivity

Consider N adaptively coupled phase oscillators [25], [29], [38]

$$\dot{\phi}_i = \omega_i + \sum_{j=1}^{N} a_{ij}\kappa_{ij} f(\phi_i - \phi_j), \quad (2)$$

$$\dot{\kappa}_{ij} = -\epsilon \left[\kappa_{ij} + g(\phi_i - \phi_j)\right], \quad (3)$$

where $\phi_i \in [0, 2\pi)$ represents the phase of the ith oscillator ($i = 1, \dots, N$), ω_i is its natural frequency, and κ_{ij} is the coupling weight of the connection from node j to i. Further, f and g are 2π-periodic functions where f is the coupling function and g is the adaptation rule, and ϵ is the adaptation rate that is usually chosen to be small ($\epsilon \ll 1$). The entries a_{ij} of the adjacency matrix A describe again the connectivity of the network.

In order to find the relation between (1) and (2)–(3), we first write (1) in the form

$$\dot{\phi}_i = \omega_i + \psi_i, \quad (4)$$

$$\dot{\psi}_i = -\frac{\gamma}{M}\left[\psi_i + \frac{\sigma}{\gamma}\sum_{j=1}^{N} a_{ij}\sin(\phi_i - \phi_j + \alpha)\right], \quad (5)$$

where ψ_i is the deviation of the instantaneous phase velocity from the natural frequency ω_i. We observe that this is a system of N phase oscillators (4) augmented by the adaptation (5) of the frequency deviation ψ_i. Note that the coupling between the phase oscillators is realized in the frequency adaptation which is different from the classical Kuramoto system [39]. As we know from the theory of adaptively coupled phase oscillators [25], [29], a frequency adaptation can also be achieved indirectly by a proper adaptation of the coupling matrix.

In order to introduce coupling weights into system (4)–(5), we express the frequency deviation ψ_i as the sum $\psi_i = \sum_{j=1}^{N} a_{ij}\chi_{ij}$ of the dynamical power flows χ_{ij} from the nodes j that are coupled with node i. The power flows are governed by the equation $\dot{\chi}_{ij} = -\epsilon\left(\chi_{ij} + g(\phi_i - \phi_j)\right)$, where $g(\phi_i - \phi_j) \equiv \sigma\sin(\phi_i - \phi_j + \alpha)/\gamma$ are their stationary values [17] and $\epsilon = \gamma/M$. It is straightforward to check that ψ_i, defined in such a way, satisfies the dynamical equation (5).

As a result, we have shown that the swing equation (4)–(5) can be written as the following system of adaptively coupled phase oscillators

$$\dot{\phi}_i = \omega_i + \sum_{j=1}^{N} a_{ij}\chi_{ij}, \quad (6)$$

$$\dot{\chi}_{ij} = -\epsilon\left[\chi_{ij} + g(\phi_i - \phi_j)\right]. \quad (7)$$

The obtained system corresponds to (2)–(3) with coupling weights χ_{ij} and coupling function $f(\phi_i - \phi_j) \equiv 1$. The coupling weights form a pseudo coupling matrix χ. Note that the base network topology a_{ij} of the phase oscillator system with inertia (1) is unaffected by the transformation.

With the introduction of the pseudo coupling weights χ_{ij}, we embed the $2N$ dimensional system (4)–(5) into a higher dimensional phase space. In [36], it was shown that the dynamics of the higher dimensional system (6)–(7) is completely governed by the system (4)–(5) on a $2N$ dimensional invariant submanifold, thereby establishing a mathematically rigorous relation.

Let us discuss the physical meaning of the coupling weights χ_{ij}. For this, we consider the power flows F_{ij} from node j to node i given by $F_{ij} = -g(\phi_i - \phi_j)$ [17]. Then each χ_{ij} is driven by the power flow from j to i. In particular, for constant F_{ij}, $\chi_{ij} \to F_{ij}$ asymptotically as $t \to \infty$ on the timescale $1/\epsilon$. Therefore, χ_{ij} acquires the meaning of a dynamical power flow.

The obtained result suggests that the power grid model is a specific realization of adaptive neural networks. Indeed, in the following, we proceed one step further and show that more complex models for synchronous machines can be represented as adaptive network as well.

B. Swing equation with voltage dynamics as adaptive network with metaplasticty

Here we generalize the results of the previous subsection for the swing equation with voltage dynamics [18], [34]:

$$M\ddot{\phi}_i + \gamma\dot{\phi}_i = P_i + \sum_{j=1}^{N} E_i E_j a_{ij} h(\phi_i - \phi_j), \quad (8)$$

$$m_i \dot{E}_i = -E_i + E_{f,i} + \sum_{j=1}^{N} a_{ij} E_j v(\phi_i - \phi_j), \quad (9)$$

where the additional dynamical variable E_i is the voltage amplitude. The functions h and v are 2π-periodic, and m_i and $E_{f,i}$ are machine parameters [18], [34]. All other variables and parameters are as in (1).

Equations (8)–(9) can be rewritten as an adaptive network (6)–(7) supplemented by (9) where $g(\phi) \equiv -E_i E_j h(\phi)/\gamma$ and $\epsilon = \gamma/M$. For this, in analogy to Section IIIA, we write (8)–(9) as

$$\dot{\phi}_i = \omega_i + \sum_{j=1}^{N} a_{ij}\chi_{ij}, \quad (10)$$

$$\dot{\chi}_{ij} = -\frac{1}{M_i}\left[\gamma\chi_{ij} - E_i E_j h(\phi_i - \phi_j)\right], \quad (11)$$

$$m_i \dot{E}_i = -E_i + E_{f,i} + \sum_{j=1}^{N} a_{ij} E_j v(\phi_i - \phi_j), \quad (12)$$

where we introduce the coordinate changes $\chi_{ij} \to \chi_{ij} + P_i/\gamma$, $E_i \to E_i + E_{f,i}$ and set $\omega_i = P_i/\gamma$. Due to the voltage dynamics (9), the adaptation function $g(\phi) = E_i(t)E_j(t)h(\phi)$ in (11) possesses additional adaptivity. This kind of meta-adaptivity (meta-plasticity) is of importance in neural networks [40] as well as for neuromorphic devices [41].

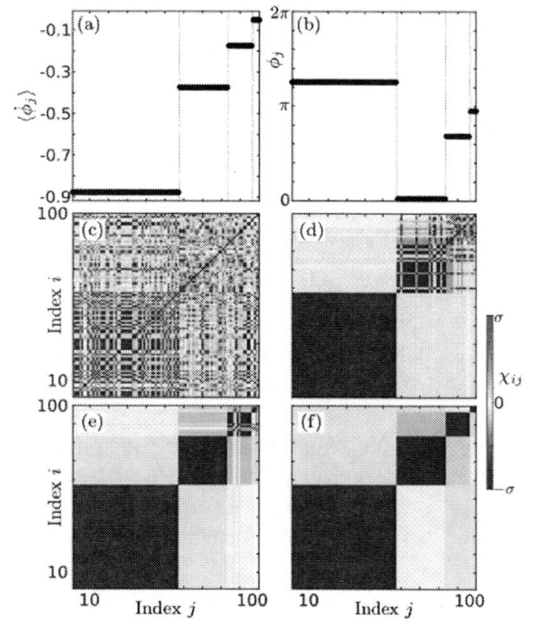

Fig. 1. Hierarchical multicluster state in a network of coupled phase oscillators with inertia. The panels (a,b) show the (temporally averaged) mean phase velocities $\langle\dot{\phi}_j\rangle$, and phase snapshots $\phi_j(t)$, respectively, at $t = 10000$. All groups of oscillators characterized by a common mean phase velocity are separated by gray dotted lines. The temporal evolution of the pseudo coupling matrix $\chi_{ij}(t)$ is presented in (c) $t = 100$, (d) $t = 1750$, (e) $t = 5000$, and (f) $t = 10000$. Starting from an incoherent state in panel (c), the largest cluster is formed first (d), and the other clusters are then successively formed depending on their size (e),(f). In (a-f) the oscillator indices are sorted in increasing order of their mean phase velocity. The state is found by numerical integration of (1) with identical oscillators $P_i = 0$ and uniform random initial conditions $\phi_i(0) \in (0, 2\pi)$, $\psi_i(0) \in (-0.5, 0.5)$. Parameters: globally coupled network $a_{ij} = 1$ for all $i \neq j$, $M = 1$, $\gamma = 0.05$, $\sigma = 0.016$, $\alpha = 0.46\pi$, $N = 100$.

IV. Mixed Frequency Cluster States in Phase Oscillator Models with Inertia

In this section, we provide a novel viewpoint of the emergence of multifrequency cluster states for phase oscillator models with inertia. In such a state all oscillators split into C groups (called clusters) each of which is characterized by a common cluster frequency Ω_μ. In particular, the temporal behavior of the ith oscillator of the μth cluster ($\mu = 1, \dots, C$) is given by $\phi_i^\mu(t) = \Omega_\mu t + \rho_i^\mu + s_i^\mu(t)$ where $\rho_i^\mu \in [0, 2\pi)$ and $s_i^\mu(t)$ are bounded functions describing different types of phase clusters characterized by the phase relation within each cluster [25]. Various types of multicluster states including the special subclass of solitary states have been extensively described for adaptively coupled phase oscillators [23], [26], [29].

In Figure 1, we present a 4-cluster state of in-phase synchronous clusters on a globally coupled network. As we know from the findings for adaptive networks [25], (hierarchical) multicluster states are built out of single cluster states whose frequency scales approximately with the number N_μ of elements in the cluster. In the zeroth-order expansion in γ, the collective cluster frequencies are given by $\Omega_\mu \approx -\sigma N_\mu \sin\alpha$. Multicluster states exist in the asymptotic limit ($\gamma \to 0$) also for networks of phase oscillators with inertia if the cluster frequencies are sufficiently different meaning the clusters are hierarchical in size. Remarkably, the pseudo coupling matrix displayed in Figure 1(f) shows the characteristic block diagonal shape that is known for adaptive networks. In particular, the oscillators within each cluster are more strongly connected than the oscillators between different clusters.

Another observation for multicluster states in networks of phase oscillators with inertia is their hierarchical emergence. As reported in [29] for adaptive networks, the clusters emerge in a temporal sequence from the largest to the smallest. In Figure 1(c-f), we show that this particular feature is also found in phase oscillators with inertia.

V. Solitary States in the German Ultra-High Voltage Power Grid

In this section, we show that multifrequency cluster states, as discussed in Figure 1, may also occur in real power grid networks, which are heterogeneous in contrast to the identical oscillators treated in the previous section. For the simulation, we consider the Kuramoto model with inertia given by (1). The network structure and the power distribution are taken from the ELMOD-DE data set provided in [42].

In Figure 2, we provide a visualization of the German ultra-high voltage power grid. In order to determine the net power consumption/generation P_i for each node in Figure 2 depicted in the inset above the map, the individual power generation and consumption at each node are compared. We obtain the net power distribution $P_i = (P_{\text{total}}/C_{\text{Total}})C_{\text{off},i} - P_{\text{off},i}$ where $P_{\text{Total}} = 36\,\text{GW}$ and $C_{\text{Total}} \approx 88.343\,\text{GW}$ are the off-peak power consumption and generation of the whole power grid network, respectively, and $C_{\text{off},i}$ and $P_{\text{off},i}$ are the off-peak power consumption and generation for each individual node, respectively. Thus power balance $\sum_i P_i = 0$ is guaranteed. For further details refer to [11], [18].

In Figure 3, we show a solitary state obtained by the simulation of model (1) with the parameters as described above for $t = 600$ and uniformly distributed random initial conditions $\phi \in (0, 2\pi)$, $\dot{\phi} \in (-1, 1)$. The temporal averages of the oscillators' phase velocities are obtained by neglecting the transient period $t \in [0, 500)$. Solitary states are special cases of multifrequency cluster states where only single nodes have a different frequency compared to the large background cluster [23]. Figure 3 (a) shows such a solitary state where 5 solitary nodes have a significantly different mean phase velocities than all the other oscillators from the large coherent cluster, which is synchronized at $\Omega_0 \approx -0.407\,\text{Hz}$. Similar results have been recently obtained in [18], [19]. Remarkably, the mean phase velocities of the solitary nodes is very close to their natural frequency, see Figure 3(b). This means that the solitary states decouple on average from the mean field of their neighborhood, i.e., $\dot{\phi}_{\text{Solitary}} = \omega_{\text{Solitary}} + \sum_j a_{ij}\chi_{ij}$ with temporal average $\langle\sum_j a_{ij}\chi_{ij}\rangle$ small compared to ω_{Solitary}.

In order to shed light on further characteristics of the solitary states, we consider the power flows, i.e, the elements of the

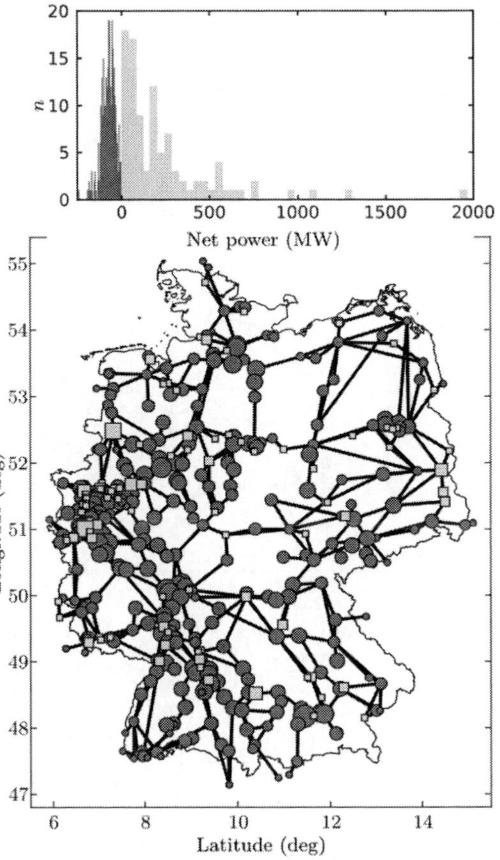

Fig. 2. Map of the German ultra-high voltage power grid consisting of 95 net generators (green squares) and 343 net consumers (red circles) connected by 662 bidirectional transmission lines (black lines). The size of each node is scaled by the actual power injected or consumed at the node from small (low absolute power) to large (high absolute power). The histogram above the map shows the net power distribution of P_i for generators (green) and consumers (red). All data are taken from the ELMOD-DE data set [42].

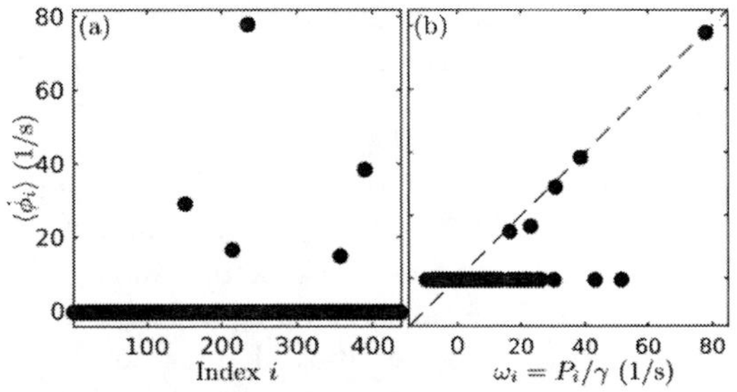

Fig. 3. Distribution of mean phase velocities for a solitary states in the German ultra-high voltage power grid. (a) Mean phase velocity $\langle\dot{\phi}_i\rangle$ for each node in the German power grid network presented in Figure 2. (b) Mean phase velocity $\langle\dot{\phi}_i\rangle$ vs. natural phase velocity $\omega_i = P_i/\gamma$ for each node in the German ultra-high voltage power grid. The dashed line shows the relation $\langle\dot{\phi}_i\rangle = \omega_i$. Parameters in model (1): $M = I\omega_G$ with $I = 40 \times 10^3$ kg m^2 and $\omega_G = 2\pi 50$ Hz, $\gamma = Ma$ with $a = 2$ Hz, $\sigma = 800$ MW, $\alpha = 0$, $t = 600$, uniformly distributed random initial conditions $\phi \in (0, 2\pi)$, $\dot{\phi} \in (-1, 1)$.

pseudo coupling matrix χ_{ij} introduced in (7). In Figure 4,

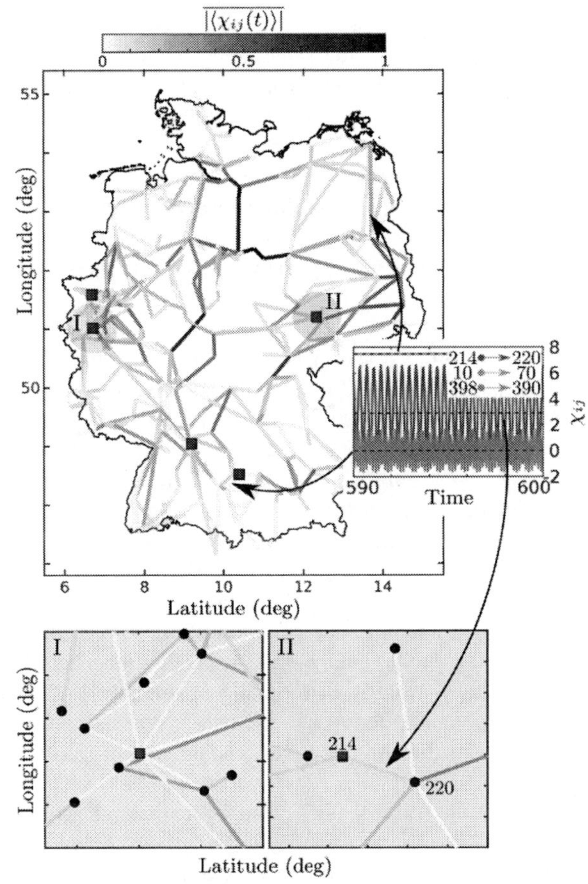

Fig. 4. Net power flow in the German ultra-high voltage power grid as given in Figure 2. The solitary nodes presented in Figure 3 are displayed as blue squares. For each transmission line, the grayscale shows the normalized average pseudo coupling weight $\overline{|\langle\chi_{ij}(t)\rangle|}$. The bar denotes the normalization for each value to the maximum for all $i, j = 1, \dots, N$. The temporal evolution is evaluated over an averaging window of 100 time units. The inset of the upper panel depicts the temporal evolution of three typical elements from the pseudo coupling matrix χ, introduced in (7), with different mean power flow levels. Arrows point to the corresponding transmission lines. The dashed black lines show the values of the average pseudo coupling weights. Panels I and II provide blow-ups of the upper panel for the solitary nodes $i = 235$ (pink shading) and $i = 214$ (light blue shading), respectively. The black nodes in the blow-ups represent the consumers and generators of the power grid network. All parameters are as in Figure 3.

we provide an overview of the pseudo coupling matrix for the results obtained in the simulation of the German ultra-high voltage power grid, see also Figure 3. Note that we do not mark the nodes of the network in Figure 4 (locations of the generators and consumers) to better visualize the characteristics of the pseudo coupling weights. We present the average coupling weights in Figure 4. As we know from the discussion in Section III, the coupling weights correspond to the dynamical power flow of each transmission line. We further know that the average value of the power flow between a solitary node and a node from the coherent cluster is small but not necessarily zero. This is in fact supported by Figure 4, see also blow-ups I and II.

The temporal variations of the power flow are presented in Figure 5. Here, only a few lines show significant temporal variations. In particular, these lines are between solitary nodes and the coherent cluster. The blow-ups support the latter observation by showing the highest values of the temporal variation of the power flow for lines from and to the solitary nodes. Besides, Figure 5 shows how far into the network power flow fluctuations are spread in the presence of solitary states. It is visible that high power fluctuations exist even between nodes of the coherent cluster. These fluctuations would not be present if all oscillators were synchronized.

The insets in the upper panels of Figures 4 and 5 depict the temporal evolution of three representative pseudo coupling weights. The three coupling weights vary periodically in time but with different amplitudes. For the coupling between two nodes of the coherent cluster ($10 \to 70$), the small variations stem from the small difference in their individual temporal dynamics which depends on their natural frequencies and the individual topological neighborhoods. In this realistic setup, the dynamical network is very heterogeneous. In contrast to the case of two nodes of the coherent clusters, the couplings between solitary nodes and a node from the coherent cluster vary much more strongly periodically in time. To understand this observation, we derive an asymptotic approximation for the dynamics of the solitary states. Using an approach similar to [25], the large power flow variations on transmission lines connecting solitary nodes can be explained. We apply a multiscale ansatz in $\epsilon = 1/K \ll 1$ to a two-cluster state. By the two-cluster state, we model the interaction of a solitary node with the coherent cluster where ϕ^1 represents the phase of the solitary node with natural frequency ω and ϕ^2 represents the phase of the coherent cluster with natural frequency Ω_0. The pseudo coupling weights between the two clusters are denoted by $\chi_{\mu\nu}$ ($\mu, \nu = 1, 2$, $\mu \neq \nu$). The ansatz reads $\phi^\mu = \Omega_\mu(\tau_0, \tau_1, \dots) + \epsilon(\phi^{(1,\mu)}) + \cdots$ and $\chi_{\mu\nu} = \chi^{(0)}_{\mu\nu} + \epsilon\chi^{(1)}_{\mu\nu} + \cdots$ with $\tau_p = \epsilon^p t$, $p \in \mathbb{N}$.

Omitting technical details, in the first order approximation in ϵ we obtain $\dot{\phi}^1 = \omega - (K/\omega^2)\cos(\omega t)$ and $\dot{\phi}^2 = \Omega_0 + (K/\omega^2)\cos(\omega t)$ [36]. Additional corrections to the oscillator frequencies appear in the third and higher orders of the expansion in ϵ and depend explicitly on ω. The latter fact is consistent with the numerical observation in Figure 3(b) that solitary nodes with a lower natural frequency may differ more strongly from their own natural frequency than the solitary oscillators with a higher natural frequency.

As we have seen, the pseudo coupling approach allows for a description of the power flow for each line individually. It shows the emergence of large power flow fluctuations at the solitary nodes, and the spreading of those fluctuations over the power grid.

VI. CONCLUSIONS

We have discussed the striking relation between phase oscillators with inertia, which are widely used for modeling power grids [8]–[22] and adaptive networks of phase oscillators, which have ubiquitous applications in physical, biological,

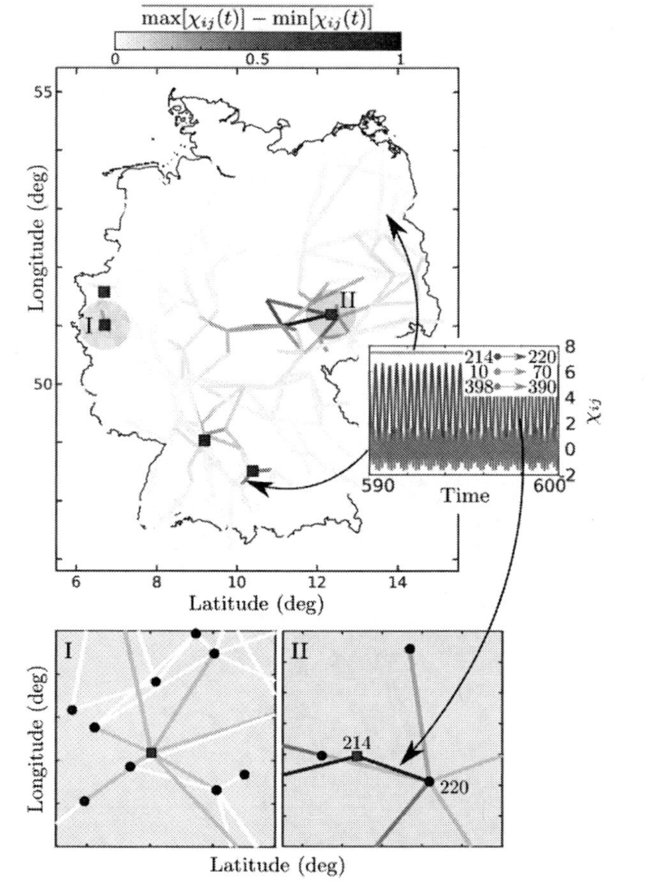

Fig. 5. Variation of the net power flow in the German ultra-high voltage power grid. The solitary nodes presented in Figure 3 are displayed as blue squares. For each transmission line, the grayscale shows the normalized amplitude of $\chi_{ij}(t)$, i.e., $\overline{\max[\chi_{ij}(t)] - \min[\chi_{ij}(t)]}$. The presentation is analogous to Figure 4. All parameters are as in Figure 3.

socioeconomic or neural systems. The introduction of the pseudo coupling matrix allows us to split the total input from all nodes into node i into power flows. Thus, the frequency deviation $\psi_i = \dot{\phi}_i - \omega_i$ in the phase oscillator model with inertia corresponds to the adaptively adjusted total input which an oscillator receives. This gives insight into the concept of phase oscillator models with inertia, which effectively takes into account the feedback loop of self-adjusted coupling with all other oscillators. Additionally, our novel theoretical framework allows for a generalization to swing equations with voltage dynamics [34].

Our first example shows that the theory of building blocks developed for adaptively coupled phase oscillators can be transferred to explain the emergence of multicluster states in networks of coupled phase oscillators with inertia. These findings are of crucial importance for studying power grid models with respect to emergent multistability and dynamical effects that lead to desynchronization [43].

In fact, a properly functioning real-world power grid should be completely synchronized, i.e., clustering into different groups with different frequencies would be undesirable. However, multicluster states can still have practical relevance, since they influence the destabilization of the synchronous state. Thus, it is important to study when they occur, in order to be able to take control measures to prevent them. For instance, recent works [18], [19] have shown that the solitary states, which are a subclass of multicluster states, arise naturally in the desynchronization transition of real-world power grid networks (German and Scandinavian power grid), and that this knowledge is essential for an efficient power grid control. For the German power grid, we provide an additional example and show analytically how the techniques developed for adaptive networks are used to characterize the emergent solitary states. We have shown that the concept of pseudo coupling weights is powerful tool to analyze the dynamical spreading of power flow fluctuations. Therefore, we believe that this concept can be used in the future to design novel detection and control approaches for modern power grid networks.

ACKNOWLEDGMENT

This work was supported by the German Research Foundation DFG, Project Nos. 411803875 and 440145547.

REFERENCES

[1] P. W. Sauer and M. A. Pai, *Power system dynamics and stability*. Upper Saddle River, NJ: Prentice hall, 1998.

[2] G. Filatrella, A. H. Nielsen, and N. F. Pedersen, "Analysis of a power grid using a Kuramoto-like model," *Eur. Phys. J. B*, vol. 61, no. 4, pp. 485–491, 2008.

[3] F. Dörfler, M. Chertkov, and F. Bullo, "Synchronization in complex oscillator networks and smart grids," *Proc. Natl. Acad. Sci. U.S.A.*, vol. 110, no. 6, pp. 2005–2010, 2013.

[4] T. Nishikawa and A. E. Motter, "Comparative analysis of existing models for power-grid synchronization," *New J. Phys.*, vol. 17, no. 1, p. 015012, 2015.

[5] S. Auer, K. Kleis, P. Schultz, J. Kurths, and F. Hellmann, "The impact of model detail on power grid resilience measures," *Eur. Phys. J. Spec. Top.*, vol. 225, no. 3, pp. 609–625, 2016.

[6] F. Dörfler and F. Bullo, "Synchronization in complex networks of phase oscillators: A survey," *Automatica*, vol. 50, no. 6, pp. 1539–1564, June 2014.

[7] F. A. Rodrigues, T. K. D. M. Peron, P. Ji, and J. Kurths, "The Kuramoto model in complex networks," *Phys. Rep.*, vol. 610, pp. 1–98, 2016.

[8] M. Rohden, A. Sorge, M. Timme, and D. Witthaut, "Self-organized synchronization in decentralized power grids," *Phys. Rev. Lett.*, vol. 109, p. 064101, August 2012.

[9] S. P. Cornelius, W. L. Kath, and A. E. Motter, "Realistic control of network dynamics," *Nat. Commun.*, vol. 4, p. 1942, 2013.

[10] A. E. Motter, S. A. Myers, M. Anghel, and T. Nishikawa, "Spontaneous synchrony in power-grid networks," *Nat. Phys.*, vol. 9, pp. 191–197, 2013.

[11] P. J. Menck, J. Heitzig, J. Kurths, and H. J. Schellnhuber, "How dead ends undermine power grid stability," *Nat. Commun.*, vol. 5, p. 3969, 2014.

[12] D. Witthaut, M. Rohden, X. Zhang, S. Hallerberg, and M. Timme, "Critical links and nonlocal rerouting in complex supply networks," *Phys. Rev. Lett.*, vol. 116, p. 138701, 2016.

[13] S. Auer, F. Hellmann, M. Krause, and J. Kurths, "Stability of synchrony against local intermittent fluctuations in tree-like power grids," *Chaos*, vol. 27, no. 12, p. 127003, September 2017.

[14] V. Mehrmann, R. Morandin, S. Olmi, and E. Schöll, "Qualitative stability and synchronicity analysis of power network models in port-Hamiltonian form," *Chaos*, vol. 28, no. 10, p. 101102, 2018.

[15] P. Jaros, S. Brezetsky, R. Levchenko, D. Dudkowski, T. Kapitaniak, and Y. Maistrenko, "Solitary states for coupled oscillators with inertia," *Chaos*, vol. 28, p. 011103, 2018.

[16] B. Schäfer, C. Beck, K. Aihara, D. Witthaut, and M. Timme, "Non-Gaussian power grid frequency fluctuations characterized by Levy-stable laws and superstatistics," *Nature Energy*, vol. 3, p. 119, 2018.

[17] B. Schäfer, D. Witthaut, M. Timme, and V. Latora, "Dynamically induced cascading failures in power grids," *Nat. Commun.*, vol. 9, p. 1975, May 2018.

[18] H. Taher, S. Olmi, and E. Schöll, "Enhancing power grid synchronization and stability through time delayed feedback control," *Phys. Rev. E*, vol. 100, no. 6, p. 062306, 2019.

[19] F. Hellmann, P. Schultz, P. Jaros, R. Levchenko, T. Kapitaniak, J. Kurths, and Y. Maistrenko, "Network-induced multistability through lossy coupling and exotic solitary states," *Nat. Commun.*, vol. 11, p. 592, January 2020.

[20] C. Kuehn and S. Throm, "Power network dynamics on graphons," *SIAM J. Appl. Dyn. Syst.*, vol. 79, no. 4, pp. 1271–1292, July 2019.

[21] C. H. Totz, S. Olmi, and E. Schöll, "Control of synchronization in two-layer power grids," *Phys. Rev. E*, vol. 102, p. 022311, 2020.

[22] F. Molnar, T. Nishikawa, and A. E. Motter, "Asymmetry underlies stability in power grids," *Nat. Commun.*, vol. 12, p. 1457, 2021.

[23] R. Berner, A. Polanska, E. Schöll, and S. Yanchuk, "Solitary states in adaptive nonlocal oscillator networks," *Eur. Phys. J. Spec. Top.*, vol. 229, no. 12-13, pp. 2183–2203, September 2020.

[24] I. V. Belykh, B. N. Brister, and V. N. Belykh, "Bistability of patterns of synchrony in Kuramoto oscillators with inertia," *Chaos*, vol. 26, p. 094822, 2016.

[25] R. Berner, E. Schöll, and S. Yanchuk, "Multiclusters in networks of adaptively coupled phase oscillators," *SIAM J. Appl. Dyn. Syst.*, vol. 18, no. 4, pp. 2227–2266, December 2019.

[26] R. Berner, J. Fialkowski, D. V. Kasatkin, V. I. Nekorkin, S. Yanchuk, and E. Schöll, "Hierarchical frequency clusters in adaptive networks of phase oscillators," *Chaos*, vol. 29, no. 10, p. 103134, October 2019.

[27] L. Tumash, S. Olmi, and E. Schöll, "Stability and control of power grids with diluted network topology," *Chaos*, vol. 29, p. 123105, 2019.

[28] S. Olmi, "Chimera states in coupled Kuramoto oscillators with inertia," *Chaos*, vol. 25, p. 123125, 2015.

[29] D. V. Kasatkin, S. Yanchuk, E. Schöll, and V. I. Nekorkin, "Self-organized emergence of multi-layer structure and chimera states in dynamical networks with adaptive couplings," *Phys. Rev. E*, vol. 96, no. 6, p. 062211, December 2017.

[30] S. Olmi, A. Navas, S. Boccaletti, and A. Torcini, "Hysteretic transitions in the Kuramoto model with inertia," *Phys. Rev. E*, vol. 90, no. 4, p. 042905, October 2014.

[31] X. Zhang, S. Boccaletti, S. Guan, and Z. Liu, "Explosive synchronization in adaptive and multilayer networks," *Phys. Rev. Lett.*, vol. 114, no. 3, p. 038701, January 2015.

[32] L. Tumash, S. Olmi, and E. Schöll, "Effect of disorder and noise in shaping the dynamics of power grids," *Europhys. Lett.*, vol. 123, p. 20001, 2018.

[33] Y. Yang and A. E. Motter, "Cascading Failures as Continuous Phase-Space Transitions," *Phys. Rev. Lett.*, vol. 199, p. 248302, December 2017.

[34] K. Schmietendorf, J. Peinke, R. Friedrich, and O. Kamps, "Self-organized synchronization and voltage stability in networks of synchronous machines," *Eur. Phys. J. Spec. Top.*, vol. 223, p. 2577, 2014.

[35] M. Ciszak, F. Marino, A. Torcini, and S. Olmi, "Emergent excitability in populations of nonexcitable units," *Phys. Rev. E*, vol. 102, no. 5, p. 050201(R), November 2020.

[36] R. Berner, S. Yanchuk, and E. Schöll, "What adaptive neuronal networks teach us about power grids," *Phys. Rev. E*, vol. 103, p. 042315, 2021.

[37] H. Sakaguchi and Y. Kuramoto, "A soluble active rotater model showing phase transitions via mutual entertainment," *Prog. Theor. Phys*, vol. 76, no. 3, pp. 576–581, 1986.

[38] R. Berner, S. Vock, E. Schöll, and S. Yanchuk, "Desynchronization transitions in adaptive networks," *Phys. Rev. Lett.*, vol. 126, no. 2, p. 028301, January 2021.

[39] Y. Kuramoto, *Chemical Oscillations, Waves and Turbulence*. Berlin: Springer-Verlag, 1984.

[40] W. C. Abraham, "Metaplasticity: tuning synapses and networks for plasticity," *Nat. Rev. Neurosci.*, vol. 9, p. 387, 2008.

[41] R. A. John, F. Liu, N. A. Chien, M. R. Kulkarni, C. Zhu, Q. D. Fu, A. Basu, Z. Liu, and N. Mathews, "Synergistic gating of electro-iono-photoactive 2d chalcogenide neuristors: Coexistence of hebbian and homeostatic synaptic metaplasticity," *Adv. Mater.*, vol. 30, no. 25, p. 1800220, June 2018.

[42] J. Egerer, "Open Source Electricity Model for Germany (ELMOD-DE)," Deutsches Institut für Wirtschaftsforschung (DIW), Tech. Rep., 2016.
[43] M. Anvari, F. Hellmann, and X. Zhang, "Introduction to focus issue: Dynamics of modern power grids," *Chaos*, vol. 30, no. 6, p. 063140, June 2020.

Pointing out the Convolution Problem of Stochastic Aggregation Methods for the Determination of Flexibility Potentials at Vertical System Interconnections

Johannes Gerster
and Sebastian Lehnhoff
Department of Computing Science
CvO Universität Oldenburg
Email: johannes.gerster@uol.de

Marcel Sarstedt
and Lutz Hofmann
Institute of Electric Power Systems
Leibniz Universität Hannover
Email: sarstedt@ifes.uni-hannover.de

Eric MSP Veith
OFFIS e.V.
R&D Division Energy
Oldenburg, Germany
Email: eric.veith@offis.de

***Abstract*—The increase of generation capacity in the area of responsibility of the Distribution System Operator requires strengthening of coordination between Transmission System Operator and Distribution System Operator in order to prevent conflicting or counteracting use of flexibility options. For this purpose, methods for the standardized description and identification of the aggregated flexibility potential of distribution grids are developed in the context of Smart Grids management and control. Approaches for identifying the flexible operation region of distribution grids can be categorized into two main classes: Data-driven/stochastic approaches and optimization-based approaches. While the latter have the advantage of working in real-world scenarios where no full grid models exist, when relying on naiv̈e sampling strategies, they suffer from poor coverage of the edges of the feasible operation region. To underpin the need for improved sampling strategies for data-driven approaches, in this paper we point out and analyse the shortcomings of naiv̈e sampling strategies with focus on the problem of leptocurtic distribution of resulting interconnection power flows. We refer to this problem as *convolution problem*, as it can be traced back to the fact that the probability density function of the sum of two or more independent random variables is the convolution of their respective probability density functions. To demonstrate the *convolution problem*, we construct a series of synthetic 0.4 kV feeders, which are characterized by an increasing number of nodes and apply a sampling strategy to them that draws set-values for the controllable distributed energy resources from independent uniform distributions. By calculating the power flow for each sample in each feeder, we end up with a collapsing point cloud of interconnection power flows clearly indicating the *convolution problem*.**

Keywords*—TSO/DSO-coordination; convolution of probability distributions; aggregation of flexibilities; feasible operation region; hierarchical grid control.

I. INTRODUCTION

The increasing share of Distributed Energy Resources (DERs) in the electrical energy system leads to new challenges for both, Transmission System Operator (TSO) and Distribution System Operator (DSO). Flexibility services for congestion management and balancing, so far mostly provided by large scale thermal power plants directly connected to the Transmission Grid (TG), increasingly have to be provided by DERs connected to the Distribution Grid (DG). Thus, DGs evolve from formerly mostly passive systems to Active Distribution Grids (ADGs) that contain a variety of controllable components interconnected via communication infrastructure and whose dynamic behaviour is characterized by higher variability of power flows and greater simultaneity factors.

TSO-DSO coordination is an important topic which has been pushed by ENTSO-E during the last years [1]–[4]. Coordination between grid operators has to be strengthened to prevent conflicting or counteracting use of flexibility options [5]. To reduce complexity for TSOs at the TSO/DSO interface and to enable TSOs to consider the flexibility potential of DGs in its operational management and optimization processes, methods are needed which allow for the determination and standardized representation of the aggregated flexibility potential of DGs.

The aggregated flexibility potential of a DG can be described as region in the PQ-plane that is made up from the set of feasible Interconnection Power Flows (IPFs) [6]. Thereby, feasible IPFs are IPFs which can be realized by using the flexibilities of controllable DERs and controllable grid components such as On-Load Tap Changer (OLTC) transformers in compliance with grid constraints i.e., voltage limits and maximum line currents.

In the literature, there are various concepts to determine the Feasible Operation Region (FOR) of DGs. They can be categorized into two main classes: Data-driven/stochastic approaches and optimization-based approaches.

For data-driven approaches, the general procedure is such that a set of random control scenarios is generated by assigning set-values from a uniform distribution to each controllable unit. By means of load flow calculations, the resulting IPFs are determined for each control scenario and classified into feasible IPFs (no grid constraints are violated) and non-feasible IPFs (at least one grid constraint is violated). The resulting point cloud of feasible IPFs in the PQ-plane serves as stencil for the FOR [6]. A problem that comes with this approach is that drawing set-values from independent uniform distributions leads to an unfavourable distribution of the resulting IPFs in the PQ-plane and extreme points on the margins of the FOR are not captured well [7].

This is where optimization-based methods come into play. The basic idea behind these methods is not to randomly sample IPFs but systematically identify marginal IPFs by solving a series of Optimal Power Flow (OPF) problems [8]. In addition to better coverage of the FOR, optimization-based approaches have the advantage of higher computational efficiency. An important drawback is however that, except for approaches which solve the OPF heuristically, solving the underlying OPF requires an explicit grid model [9]. On the other hand, the only heuristic approach published so far, suffers from poor automatability as it relies on manual tweaking of hyperparameters [10].

In practice, considering the huge size of DGs, complete data related to grid topology (data related to operating equipment incl. its characteristics and topological connections) is not commonly stored [11], which complicates parametrization of explicit grid models. In such circumstances, black-box machine learning (ML) models trained on measurement data provided by current smart meters can be an alternative to physics-based, explicit grid models [12]. Due to their compatibility with black-box grid models, we argue that it is worthwhile to research and improve data-driven approaches to determine the FOR of DGs. This paper is intended to point out and analyse the problem of leptocurtic distribution of IPFs with naïve sampling strategies and thus to underpin the need for more effective and more efficient data-driven sampling strategies, such as those published by Contreras *et al.* [13], when this paper was already advanced.

The remainder of the paper is structured as follows: A survey on existing approaches (data-driven and optimization-based) and the contribution of this paper are given in Section II. Next, in Section III the construction of a series of synthetic feeders with increasing number of nodes is explained. On the basis of these feeders, we performed our sampling experiments, whose results are presented in Section IV. Finally, in Section V the paper is summarized, the conclusion is given and an idea for a distributed sampling strategy is presented which does not suffer from unfavourable distribution of resulting IPFs.

II. Survey on Grid Flexibility Aggregation Methods and Contribution of this Paper

As outlined in the introduction, relevant literature can be grouped into two main categories: Data-driven/stochastic and optimization-based approaches for exploring the FOR of DGs.

A. *Data-driven approaches*

Heleno *et al.* [7] are the first to come up with the idea of estimating the flexibility range in each primary substation node to inform the TSO about the technically feasible aggregated flexibility of DGs. In order to enable the TSO to perform a cost/benefit evaluation, they also include the costs associated with adjusting the originally planned operating point of flexible resources in their algorithm. In the paper two variants of a Monte Carlo simulation approach are presented, which differ in the assignment of set-values to the flexible resources. While in the first approach independent random set-values for changing active and reactive power injection are associated to each flexible resource, in the second approach a negative correlation of one between generation and load at the same bus was considered. In a direct comparison of the two presented approaches, the approach with negative correlation between generation and load at the same bus performs better and results in a wider flexibility range and lower flexibility costs with a smaller sample size. Nevertheless, even with this approach, the capability to find marginal points in the FOR is limited. Therefore, in the outlook the authors suggest the formulation of an optimization problem in order to overcome the limitations of the Monte Carlo simulation approach, increasing the capability to find extreme points of the FOR and reducing the computational effort. In Silva *et al.* [8], which is discussed in the next subsection, the authors take up this idea again.

Mayorga Gonzalez *et al.* [6] extend in their paper the methodology presented by Heleno *et al.* [7]. First, they describe an approach to approximate the FOR of an ADG for a particular point in time, assuming that all influencing factors are known. For this, they use the first approach of Heleno *et al.* [7] for sampling IPFs (the one that does not consider correlations). That is, random control scenarios are generated by assigning set-values from independent uniform distributions to all controllable units. In contrast to [7], no cost values are calculated for the resulting IPFs. Instead, for describing the numerically computed FOR with sparse data, the region is approximated with a polygon in the complex plane. In addition, a probabilistic approach to assess in advance the flexibility associated to an ADG that will be available in a future time interval under consideration of forecasts which are subject to uncertainty is proposed. The authors mention that for practical usage the computation time for both approaches has to be significantly reduced. However, the problem of unfavourable distribution of the resulting IPF point cloud, when drawing control scenarios from independent uniform distributions, which is a mayor factor for the low computational efficiency, is not discussed.

When this paper was already advanced, Contreras *et al.* [13] came up with new sampling methods for data-driven approaches. They show that, when focusing the vertices of the flexibility chart of flexibility providing units during sampling, the quality of the data-driven approach can be dramatically improved in comparison to the naiv̈e sampling. On top of that, they present a comparison of OPF-based and data-driven approaches, whose results show that with their improved sampling strategies both approaches are capable of assessing the FOR of radial distribution grids. But for grids with large number of buses, OPF-based methods are still better suited.

B. *Optimization-based approaches*

Silva *et al.* [8] address the main limitation of their sampling-based approach in Heleno *et al.* [7], namely estimating extreme points in the FOR. To this end, they propose a methodology which is based on formulating an optimization problem with below-mentioned objective function, whose minimization for different ratios of α and β allows to capture the perimeter of the flexibility area.

$$\alpha\, P_{DSO \to TSO} + \beta\, Q_{DSO \to TSO} \qquad (1)$$

where $P_{DSO \to TSO}$ and $Q_{DSO \to TSO}$ are the active and reactive power injections at the TSO-DSO boundary nodes. Silva *et al.* [8] work out that the underlying optimization problem represents an OPF problem. Due to its robust characteristics, they use the primal-dual, a variant of the interior point methods to solve it. The methodology was evaluated in simulation and validated in real field-tests on MV distribution networks in France. The comparison of simulation results with the random sampling algorithm in Heleno *et al.* [7] shows the superiority of the optimization-based approach by illustrating its capability to identify a larger flexibility area and to do it within a shorter computing time.

Capitanescu [14] proposes the concept of active-reactive power (PQ) chart, which characterizes the short-term flexibility capability of active distribution networks to provide ancillary services to TSO. To support this concept, an AC optimal power flow-based methodology to generate PQ capability charts of desired granularity is proposed and illustrated in a modified 34-bus distribution grid.

Contreras *et al.* [9] present a linear optimization model for the aggregation of active and reactive power flexibility of distribution grids at a TSO-DSO interconnection point. The power flow equations are linearized by using the Jacobian matrix of the Newton-Raphson algorithm. The model is complemented with non-rectangular linear representations of typical flexibility providing units, increasing the accuracy of the distribution grid aggregation. The obtained linear programming system allows a considerable reduction of the required computing time for the process. At the same time, it maintains the accuracy of the power flow calculations and increases the stability of the search algorithm while considering large grid models.

Fortenbacher *et al.* [15] present a method to compute reduced and aggregated distribution grid representations that provide an interface in the form of active and reactive power (PQ) capability areas to improve TSO-DSO interactions. Based on a lossless linear power flow approximation, they derive polyhedral sets to determine a reduced PQ operating region capturing all voltage magnitude and branch power flow constraints of the DG. While approximation errors are reasonable, especially for low voltage grids, computational complexity is significantly reduced with this method.

Sarstedt *et al.* [10] provide a detailed survey on stochastic and optimization based methods for the determination of the FOR. They come up with a comparison of different FOR determination methods with regard to quality of results and computation time. For their comparison, they use the Cigré medium voltage test system. On top of that, they present a Particle Swarm Optimization (PSO) based method for FOR determination.

Contributions of this paper

In summary, it can be stated that optimization-based approaches show high computational efficiency with good coverage of the FOR. However, methods used for solving the underlying OPF problem rely—except for heuristic approaches, which have other drawbacks—on explicit grid models of the DG, which must be parametrized with grid topology data often not available in practice. Data-driven approaches, on the other hand, do not require explicit grid modeling and are compatible with black-box grid models, but suffer from low computational efficiency and poor coverage of peripheral regions of the FOR, when using conventional sampling strategies.

This is where our approach comes in. We are heading towards improved sampling strategies for data-driven approaches, which mitigate the weak points of data-driven methods (low computational efficiency and poor coverage of FOR) while retaining their advantage of being compatible with black-box grid models. As a basis for this, in this paper we are the first to come up with an experiment setup by means of which the problem of resulting convoluted distribution of IPFs with naïve sampling strategies can be analyzed and pointed out in an easily reproducible manner.

III. Experiment Setup

To show the shortcomings of naïve sampling strategies, we apply a sampling strategy that draws set-values for the controllable DERs from indpendent uniform distributions to a series of synthetic $0.4\,\mathrm{kV}$ feeders as shown in Figure 1. The feeders are characterized by an increasing number of nodes. To be able to consider the effect of the number of nodes on the IPF-sample as isolated as possible, both, the total installed power and the average transformer-node distance are chosen equal for all feeders. The installed power is distributed equally among all connected DERs:

$$P^i_{inst,DER_j} = \frac{P^i_{inst,DERs}}{N^i}, \tag{2}$$

where P^i_{inst,DER_j} is the installed power of the DER connected to the jth node n^i_j of feeder i, $P^i_{inst,DERs}$ is the total installed power of feeder i and N^i is the number of nodes of feeder i. Nodes are equally distributed along feeders as shown in Figure 1 and the line length between adjacent nodes l^i_l of feeder i with length l^i_f is:

$$l^i_l = \frac{l^i_f}{N^i}. \tag{3}$$

The transformer-node distance of node n^i_j is:

$$d^i_{t,n_j} = l^i_l \cdot j. \tag{4}$$

With (4) the average transformer-node distance $\overline{d^i_{t,n}}$ of feeder i can be written as:

$$\begin{aligned} \overline{d^i_{t,n}} &= \frac{1}{N^i}\sum_{j=1}^{N^i} d^i_{t,n_j} \\ &= \frac{1}{N^i}\sum_{j=1}^{N^i} l^i_l \cdot j = \frac{l^i_l}{N^i}\sum_{j=1}^{N^i} j \\ &= \frac{l^i_l}{N^i} \cdot \frac{N^i_j \cdot (N^i_j + 1)}{2}. \end{aligned} \tag{5}$$

Resolved after the line length l^i_l, the result is:

$$l^i_l = \overline{d^i_{t,n}} \cdot \frac{2}{N^i + 1}. \tag{6}$$

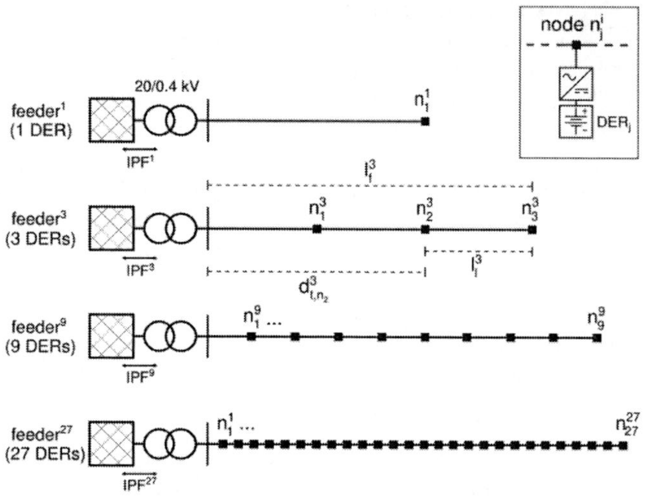

Figure 1. Synthetic 0.4 kV feeders

With (3) and (6) the length of feeder i results in:

$$l_f^i = \overline{d_{t,n}^i} \cdot \frac{2N^i}{N^i+1}. \tag{7}$$

For our experiments, we have constructed four synthetic 0.4 kV feeders. The feeders differ in the number of nodes N^i, which has been set to 1, 3, 9 or 27, respectively. Line length l_l^i and feeder length l_f^i have then been calculated according to (6) and (7). There is one DER connected to each node and the installed power $P_{inst,DERs}^i$ is distributed evenly among the DERs according to (2). To be able to cover the entire flexibility area of the feeders including its border areas where voltage band violations and/or line overloadings can be observed, all DERs are inverter-connected battery storages because they offer maximum flexibility with regard to both, active and reactive power provision. The dimensioning of the inverters has been chosen in such a way that a power factor $\cos\phi$ of 0.9 can be kept, when the maximum active power is delivered:

$$|S|_{max,DER_j}^i = \frac{P_{inst,DER_j}^i}{\cos\phi} = \frac{P_{inst,DER_j}^i}{0.9}. \tag{8}$$

Active and reactive power ranges of the battery storages are thus:

$$\left[P_{min,DER_j}^i, P_{max,DER_j}^i\right] = \left[-P_{inst,DER_j}^i, P_{inst,DER_j}^i\right]$$
$$\left[Q_{min,DER_j}^i, Q_{max,DER_j}^i\right] = \left[-|S|_{max,DER_j}^i, |S|_{max,DER_j}^i\right]. \tag{9}$$

Values for the technical parameters of the four feeders including connected DERs are listed in Table I.

For all feeders, we conduct the sampling in such a way that for each DER and each sample element we independently draw real and reactive power values from uniform distributions:

$$\begin{aligned} \mathcal{X}_{P,DER_j}^i &\sim \mathcal{U}_j^i\left[P_{min,DER_j}^i, P_{max,DER_j}^i\right] \\ \mathcal{X}_{Q,DER_j}^i &\sim \mathcal{U}_j^i\left[Q_{min,DER_j}^i, Q_{max,DER_j}^i\right]. \end{aligned} \tag{10}$$

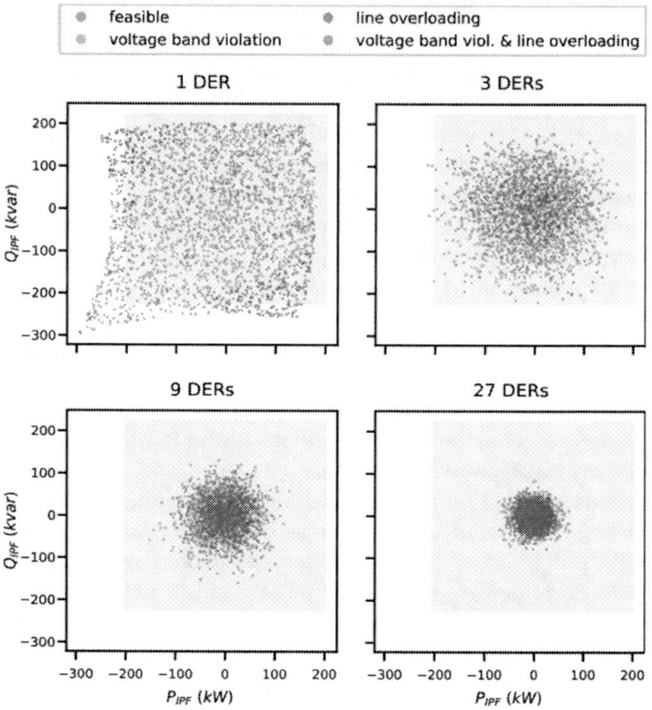

Figure 2. Results of naïve sampling strategy classified by feasibility with regard to grid constraints (voltage band limits and max. line loading); inverter constraints are neglected

After assigning active and reactive power values to each DER, the *pandapower* library [16] calculates the power flow. This way we generate a sample of size 2500 for each feeder.

Following this, the sample elements are first classified with regard to their adherence to grid constraints and in case of non-adherence with regard to the type of grid constraint violation (i.e., voltage band violation, line overload, or both). Second, inverter constraints are taken into account and the sample elements are classified with regard to adherence to both, grid and inverter constraints. In this case, sample elements are only classified as feasible, if neither grid constraints nor device constraints for any of the connected inverters occur. In case of non-feasibility, we distinguish depending on the type of constraint violation (i.e., grid constraint violation, inverter constraint violation, or both).

Finally, we plot the classification results in the domain of active and reactive IPFs P_{IPF} and Q_{IPF}.

IV. Experiment Results

The resulting plots are shown in Figure 2 and 3. The sampling has been performed once for each feeder from Figure 1. This means that Figure 2 and 3 only differ in how the sample elements are classified. While for Figure 2 only grid constraints have been considered, Figure 3 also incorporates inverter constraints. Both figures consist of four subplots—one for each of the four feeders from Figure 1. Each dot of the point clouds represents one sample element—so every subplot contains 2500

TABLE I. CONFIGURATION OF THE SYNTHETIC FEEDERS

# DERs	P_{inst,DER_j} (kW)	$\lvert S\rvert_{max,DER_j}$ (kVA)	Feeder Length (m)	Line Length (m)	Line Type	Voltage Band (pu)	Trafo Type
1	200.0	222.2	400	400	NAYY 4x150 SE	0.9–1.1	0.4 MVA 20/0.4 kV
3	66.7	74.1	600	200	NAYY 4x150 SE	0.9–1.1	0.4 MVA 20/0.4 kV
9	22.2	24.7	720	80	NAYY 4x150 SE	0.9–1.1	0.4 MVA 20/0.4 kV
27	7.4	8.2	771	29	NAYY 4x150 SE	0.9–1.1	0.4 MVA 20/0.4 kV

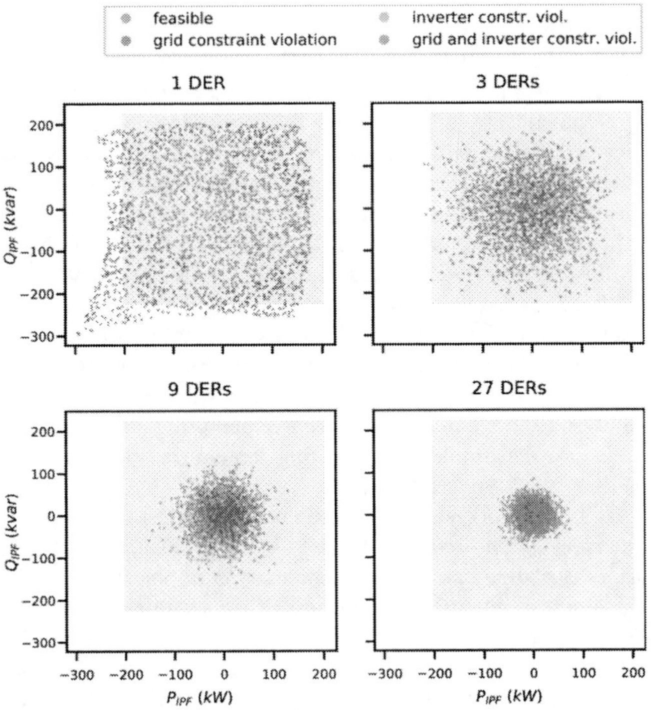

Figure 3. Results of naïve sampling strategy classified by feasibility with regard to both, grid and inverter constraints; please note: green dots are plotted above orange dots, i.e., no green dots are covered by orange dots

dots. The shaded grey area marks the theoretically known aggregated power limit for the DERs:

$$\left[P^i_{min,DERs}, P^i_{max,DERs}\right] = \left[-P^i_{inst,DERs}, P^i_{inst,DERs}\right]$$
$$\left[Q^i_{min,DERs}, Q^i_{max,DERs}\right] = \left[-|S|^i_{max,DERs}, |S|^i_{max,DERs}\right]. \quad (11)$$

For the 1 DER case, shape and structure of the point cloud look as one would expect from the configuration. It largely covers the grey area—only slightly skewed and shifted towards lower active and reactive power values, which results from active and reactive power consumption of grid elements (lines and transformer). However, when increasing the number of DERs, the *convolution problem* becomes obvious. With 3 DERs, the feasible area is still covered to some extent, but the point density already decreases strongly towards the edges. In case of 9 DERs, the point density in the edges has decreased to such an extent that hardly any sample elements are detected which show grid constraint violations. Finally, with 27 DERs the point cloud has collapsed to a fraction of the grey area and only a small part of the theoretical FOR is covered.

From the 3 DERs subplot in Figure 2, it can be seen that with increasing number of DERs not only the region covered by the sample collapses, but at the same time the border between feasible and infeasible elements (with regard to grid constraints) becomes less distinct: The absence of a sharp border between feasible and non-feasible IPFs complicates the use of multi-class classification for identifying the FOR from the sample and indicates the use of a one-class classifier for that purpose.

Figure 3, which additionally considers inverter constraints, shows an other problem of the naiv̈e sampling approach: In this consideration, not only the total area covered by the sample decreases, but also the share of feasible examples shrinks sharply, such that with 27 DERs only very few sample elements are identified which violate neither grid nor inverter constraints.

This is because with the naiv̈e sampling approach power values are assigned to each DER at once. After that, the power flow calculation is performed and only at the very end the feasiblity with regard to grid and inverter constraints is checked. Even if the constraints of only a single inverter are violated, the example is classified as non-feasible with regard to inverter constraints. If, for example, for a single converter one third of the possible power setpoints violate constraints, the likelihood to observe no constraint violations with N inverters amounts to

$$\left(1 - \frac{1}{3}\right)^N.$$

For $N = 27$ inverters this would amount to approximately 1.76×10^{-5}.

One way to address this would be to perform a successive sampling as proposed by Bremer *et al.* [17] for the use case of active power planning. With successive sampling the evaluation of inverter constraints is done immediately after the assignment of setpoints to single DERs and in case of non-feasibility drawing of setpoints is repeated until a valid configuration is found. The power flow calculation would then be carried out only after setpoints compatible with inverter constraints have been found for each DER.

To illustrate the *convolution problem*, in Figure 4 we plot the frequency density of active IPFs resulting from our sampling against the Probability Density Function (PDF) of the Bates distribution. The Bates distribution is the continuous probability distribution of the mean of n independent uniformly distributed

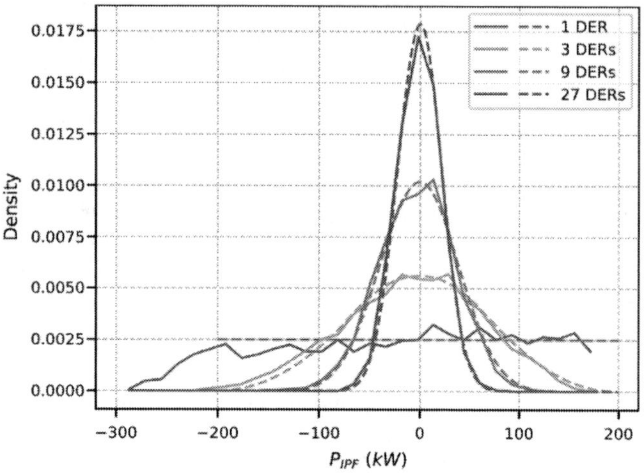

Figure 4. Frequency density of active IPFs resulting from our experiments (solid lines) compared with probability density function of Bates distribution on the interval $[-P^i_{inst,DERs}, P^i_{inst,DERs}]$ (dashed lines)

random variables on the unit interval and thus closely related to the Irwin-Hall distribution, which describes the sum of n independent uniformly distributed random variables. More general, in statistics the probability distribution of the sum of two or more independent random variables is the convolution of their individual distributions. For the variant of the Bates distribution generalized to arbitrary intervals $[a, b]$:

$$\mathcal{X}_{(a,b)} = \frac{1}{n}\sum_{k=1}^{n}\mathcal{U}_k(a,b) \tag{12}$$

this results in the following equation defining the PDF:

$$f(x) = \begin{cases} \sum_{k=0}^{n}\left[(-1)^k\binom{n}{k}\left(\frac{x-a}{b-a}-\frac{k}{n}\right)^{n-1} \operatorname{sgn}\left(\frac{x-a}{b-a}-\frac{k}{n}\right)\right] & \text{if } x \in [a,b] \\ 0 & \text{otherwise.} \end{cases} \tag{13}$$

Comparison with the Bates distribution is motivated by the fact that with equations (2), (9) and (10) the active power range from which we draw values during the sampling can be written as follows:

$$\begin{aligned}\mathcal{X}^i_{P,DER_j} &\sim \mathcal{U}^i_j\left[\frac{-P^i_{inst,DERs}}{N^i}, \frac{P^i_{inst,DERs}}{N^i}\right] \\ &= \frac{1}{N^i}\mathcal{U}^i_j\left[-P^i_{inst,DERs}, P^i_{inst,DERs}\right].\end{aligned} \tag{14}$$

For a single sample element the active IPF P^i_{IPF} is made up of the sum of active power injections of connected DERs $P^i_{DER_j}$ and the grid losses P^i_{loss}:

$$P^i_{IPF} = \sum_{j=1}^{N_i} P^i_{DER_j} + P^i_{loss} \tag{15}$$

We are interested in the distribution $\mathcal{X}^i_{P,IPF}$ of active IPFs. Ignoring grid losses P^i_{loss} in (15), with equations (14) and (15) we can write:

$$\begin{aligned}\mathcal{X}^i_{P,IPF} &\sim \sum_{j=1}^{N_i}\frac{1}{N^i}\mathcal{U}^i_j\left[-P^i_{inst,DERs}, P^i_{inst,DERs}\right] \\ &= \frac{1}{N^i}\sum_{j=1}^{N_i}\mathcal{U}^i_j\left[-P^i_{inst,DERs}, P^i_{inst,DERs}\right],\end{aligned} \tag{16}$$

which is exactly the Bates distribution on the interval $[-P^i_{inst,DERs}, P^i_{inst,DERs}]$.

V. Conclusion and Future Work

Aggregating the flexibility potential of DGs is an important prerequisite for effective TSO-DSO coordination in electric power systems with high share of generation located in the DG level. In this paper, we first gave an overview of existing flexibility aggregation methods and categorized them in terms of whether they are data-driven/stochastic or optimization-based. Following this, we discussed the strengths and weaknesses of both approaches (stochastic and optimization-based) and motivated the investigation of improved sampling strategies for data-driven approaches. As a basis for this, we presented an experimental setup by means of which we demonstrated and analyzed the shortcomings of naïve sampling strategies with focus on the problem of resulting leptokurtic distribution of IPFs.

In future work we will investigate approaches for mitigating the *convolution problem*. One idea is to formulate the sampling as a distributed optimization problem whose objective function takes into account the uniformity of the resulting set of IPFs. First experiments in this direction with the Combinatorial Optimization Heuristic for Distributed Agents (COHDA) protocol by Hinrichs *et al.* [18] and with Ripleys-K as metric for the evaluation of the distribution show promising results.

Additionally, we are working on making OPF-based methods compatible with black-box grid models by solving the underlying OPF with the help of evolutionary algorithms such as the covariance matrix adaptation evolution strategy (CMA-ES) [19] or REvol, an algorithm which was originally developed for training artificial neural networks [20]. Furthermore, we want to investigate if the total number of required objective function evaluations can be reduced when sampling the border of the FOR in one run by dynamically adapting the underlying objective function.

Acknowledgements

This work was funded by the Deutsche Forschungsgemeinschaft (DFG, German Research Foundation) – 359921210.

References

[1] ENTSO-E, "General guidelines for reinforcing the cooperation between TSO and DSO," Tech. Rep., 2015.
[2] ——, "Towards smarter grids: Developing TSO and DSO roles and interactions for the benefit of consumers," Position Paper, 2015.
[3] ——, "Distributed flexibility and the value of TSO/DSO cooperation – Fostering active customer participation to value their services on the market – An ENTSO-E position," Policy Paper, Dec. 2017.

[4] ——, "TSO-DSO Report – An integrated approach to active system management with the focus on TSO-DSO coordination in congestion management and balancing," Tech. Rep., 2019.

[5] M. Sarstedt *et al.*, "Standardized evaluation of multi-level grid control strategies for future converter-dominated electric energy systems," *at - Automatisierungstechnik*, vol. 67, no. 11, pp. 936–957, Nov. 2019.

[6] D. Mayorga Gonzalez *et al.*, "Determination of the time-dependent flexibility of active distribution networks to control their TSO-DSO interconnection power flow," in *2018 Power Systems Computation Conference (PSCC)*. Dublin, Ireland: IEEE, Jun. 2018, pp. 1–8.

[7] M. Heleno *et al.*, "Estimation of the flexibility range in the transmission-distribution boundary," in *2015 IEEE Eindhoven PowerTech*. Eindhoven, Netherlands: IEEE, Jun. 2015, pp. 1–6.

[8] J. Silva *et al.*, "Estimating the active and reactive power flexibility area at the TSO-DSO interface," *IEEE Transactions on Power Systems*, vol. 33, no. 5, pp. 4741–4750, Sep. 2018.

[9] D. A. Contreras and K. Rudion, "Improved assessment of the flexibility range of distribution grids using linear optimization," in *2018 Power Systems Computation Conference (PSCC)*. Dublin, Ireland: IEEE, Jun. 2018, pp. 1–7.

[10] M. Sarstedt, L. Kluß, J. Gerster, T. Meldau, and L. Hofmann, "Survey and comparison of optimization-based aggregation methods for the determination of the flexibility potentials at vertical system interconnections," *Energies*, vol. 14, no. 3, p. 687, 2021.

[11] R. Singh, E. Manitsas, B. C. Pal, and G. Strbac, "A recursive Bayesian approach for identification of network configuration changes in distribution system state estimation," *IEEE Transactions on Power Systems*, vol. 25, no. 3, pp. 1329–1336, Aug. 2010.

[12] P. Barbeiro, J. Krstulovic, F. J. Soares, H. Teixeira, and J. P. Iria, "State estimation in distribution smart grids using autoencoders," *IEEE 8th International Power Engineering and Optimization Conference (PEOCO2014)*, pp. 358–363, 2014.

[13] D. A. Contreras and K. Rudion, "Computing the feasible operating region of active distribution networks: Comparison and validation of random sampling and optimal power flow based methods," *IET Generation, Transmission & Distribution*, vol. 2021, pp. 1–13, 2021.

[14] F. Capitanescu, "TSO–DSO interaction: Active distribution network power chart for TSO ancillary services provision," *Electric Power Systems Research*, vol. 163, pp. 226–230, Oct. 2018.

[15] P. Fortenbacher and T. Demiray, "Reduced and aggregated distribution grid representations approximated by polyhedral sets," *International Journal of Electrical Power & Energy Systems*, vol. 117, 2020, 105668.

[16] L. Thurner *et al.*, "Pandapower—An open-source Python tool for convenient modeling, analysis, and optimization of electric power systems," *IEEE Transactions on Power Systems*, vol. 33, no. 6, pp. 6510–6521, Nov. 2018.

[17] J. Bremer and M. Sonnenschein, "Sampling the search space of energy resources for self-organized, agent-based planning of active power provision," in *Environmental Informatics and Renewable Energies - 27th International Conference on Informatics for Environmental Protection*, Hamburg, Sep. 2013, pp. 214–222.

[18] C. Hinrichs and M. Sonnenschein, "A distributed combinatorial optimisation heuristic for the scheduling of energy resources represented by self-interested agents," *Int. J. of Bio-Inspired Computation*, vol. 10, no. 2, pp. 69–78, Jul. 2017.

[19] N. Hansen, "The CMA Evolution Strategy: A Comparing Review," *Towards a new evolutionary computation*, pp. 75–102, 2006.

[20] E. M. Veith, B. Steinbach, and M. Ruppert, "An evolutionary training algorithm for artificial neural networks with dynamic offspring spread and implicit gradient information," in *The Sixth International Conference on Emerging Network Intelligence (EMERGING 2014)*, Rome, Aug. 2014.

Hybrid Renewable Energy System Optimization is Lacking Consideration of System Resilience and Robustness: An Overview

Lasse Hammer and Eric MSP Veith
OFFIS e.V.
Oldenburg, Germany
Emails: {lasse.hammer, eric.veith}@offis.de

Abstract—**This paper reviews existing literature that focuses on optimizing Hybrid Renewable Energy System (HRES) regarding their incorporation of resilience and robustness properties and gives an overview of commonly used techniques in the field. HRES are energy systems consisting of renewable energy sources, as well as traditional fuel based generators as backup. In the current transformation phase of energy generation, it is important to size those systems large enough but as small as possible. Today, a plethora of optimization goals and techniques, as well as approaches to model and simulate the systems are known to researchers. Since no common definition of resilience and robustness exists for cyber-physical systems like HRES, different definitions are compared and explained. The review shows that a research gap exists in taking resilience and robustness into account when optimizing HRES. An outlook on how to address this research gap using Adversarial Resilience Learning (ARL) is also given.**

Keywords—*HRES; optimization; resilience; robustness; ARL.*

I. Introduction

In order to reach the climate targets defined in the Paris Climate Agreement [1], the supply of electrical energy needs to be shifted from fossil to renewable sources. This often requires a redesign of the power grid as renewable energy sources are less dependable and therefore require ways of storing energy not necessarily needed before. At this point, the question of how to transform energy systems arises. This transformation will most likely not happen over night, which results in an intermittent state. The mix of renewable energy sources, energy storage and fossil sources as backup forms a Hybrid Renewable Energy System (HRES) [2]. Those systems can be built with a connection to the energy grid [3][4], or as standalone systems that provide electricity with low dependence on fuel in remote areas [5][6].

In order to design efficient and reliable HRES, a lot of research has been done on HRES optimization. Often, the optimization focuses on economical and technical aspects of the system like the cost of the generated energy or the system's ability to meet the energy demand. Recently, environmental and socio-political goals such as CO_2 emission of the system and impact on the local community have been scrutinized as well [7].

Renewable energy systems are often highly distributed and thus require extensive communication between components [8]. This leads to more and more digitization, effectively making the energy grid a large cyber-physical system [9], which poses different challenges. The recent blackout in the Ukraine [10] for example was caused by a cyber attack on the energy infrastructure. Other unpredictable events like the overloading of a substation in Europe in 2021, which lead to a system separation [11] or the increasing amounts of natural disasters like earthquakes, storms and floods show that energy grids can be disrupted in unforseeable ways [12][13]. In order to withstand such challenges, HRES must be resilient and robust, which begs the question whether resilience and robustness are considered in HRES optimization. Therefore, this paper gives an overview of HRES optimization, explains common techniques and reviews them regarding their incorporation of resilience and robustness.

Since no common definition of resilience and robustness exists, in Section II, we first compare existing definitions and decide on how we use the terms in this paper. In Section III, HRES components are introduced and their function is clarified. We describe optimization problems in general in Section IV, and methods used in HRES optimization in Section V. After that, in Section VI, common simulation techniques are explored. Section VII explains frequent optimization goals and reviews their consideration of resilience and robustness properties. Those findings are combined in Section VIII to identify a research gap. Finally, in Section IX a summary is given, as well as an outlook on how the research gap will be addressed.

II. Energy grid resilience and robustness

Throughout the literature exists no commonly agreed upon definition of resilience of cyber-physical systems. Arghandeh et al. [14] define cyber-physical resilience as

Definition 1. *The resilience of a system presented with an unexpected set of disturbances is the system's ability to reduce the magnitude and duration of the disruption. A resilient system downgrades its functionality and alters its structure in an agile way.*

The Presidential Policy Directive 21 [15] of the United States of America defines it as

Definition 2. *The ability to prepare for and adapt to changing conditions and withstand and recover rapidly from disruptions. Resilience includes the ability to withstand and recover from deliberate attacks, accidents, or naturally occurring threats or incidents.*

The definitions show, that resilience of cyber-physical systems is concerned with the handling of unexpected disturbances. In order to remain functional, a resilient system downgrades its functionality and recovers back to the regular operating mode quickly. In energy grids the downgrade of functionality might mean shutting down subgrids in order to keep the rest of the grid stable. Resilience often gets confused with robustness. Arghandeh et al. [14] define cyber-physical robustness as

Definition 3. *Robustness is the ability of a system to cope with a given set of disturbances and maintain its functionality.*

The main difference is that a robust system maintains it's functionality, while a resilient system can downgrade and recover it's functionality. In energy grids robustness is, e.g., achieved with control energy [16]. Control energy is used to level frequency deviations that occur when energy demand and generation are not equal.

III. Hybrid Renewable Energy System Components

Hybrid Renewable Energy Systems are built from different components. They combine renewable energy sources like photovoltaic and wind turbines with traditional fossil fuel based generators to provide energy locally without high dependence on fossil fuels [17]. Traditionally, HRES are built for specific use cases like power sources for cities [18], small villages [19] or even buildings [20]. They can exist as standalone systems to provide power in remote areas [5] [6], or can be connected to the power grid [3] [4] in areas, where electricity is available anyways. This section explains commonly used components of HRES and describes their function.

A. Photovoltaic

Photovoltaic (PV) cells generate electricity by absorbing light. They consist of semiconductor material that forms an electric field. Once light hits the cell, electrons are knocked loose from the semiconductor's atoms. The electrons flow between the positive and negative side of the electric field, which creates a current and thus electricity. PV cells are usually connected and mounted to form PV modules, which can then be installed to harvest energy [21].

B. Wind turbine

Wind turbines convert the kinetic energy of wind into electricity. They achieve that by capturing the wind energy with rotor blades that make the rotor turn. The rotor is connected to a generator that turns the kinetic energy of the rotating motion into electricity. Usually, the rotor and generator are mounted on a tower to allow for large rotor diameters and thus a higher energy output [22].

C. Battery

Batteries are a form of energy store. They can store electrical energy as chemical energy. Batteries are composed of two electrodes, called anode and cathode, that are submerged in an electrolyte. When discharging a battery, a reduction-oxidation reaction occurs at the electrodes. At the anode, electrons are set free, which flow through the electric circuit attached to the battery to the cathode, thus creating a current. To balance that, positively charged ions move from the anode through the electrolyte to the cathode. If the right electrode material is used, batteries can be recharged by attaching an electricity source to the battery, which reverses the aforementioned process. Batteries can be used to store energy created by whether dependent energy sources if supply is higher than demand [23].

D. Diesel generator

Diesel generators are diesel engines that are connected to a generator. By compressing air and combusting diesel fuel, pistons in the engine move up and down cylinders. This motion is converted into a rotation of the crankshaft via connecting rods that connect the crankshaft to the pistons. The rotation powers a generator that transforms the kinetic energy to electrical energy [24]. The generator operates on the same principle as the generator in a wind turbine. The difference between a diesel generator and a wind turbine is the creation of the rotation.

E. Gas turbine

Gas turbines burn gas to create a rotating motion. The turbine compresses an air and gas mixture with rotating blades attached to a center shaft. The mixture is then ignited and the hot gases spin blades connected to the same shaft. The rotation is used to compress air on the compressor side and to power a generator similar to the generators used in wind turbines or diesel generators. An added benefit is the usage of the hot gas mixture for heating purposes, which allows for very efficient operation of gas turbines [25].

F. Hydrogen fuel cell

Hydrogen fuel cells use hydrogen and oxygen to create electricity. Although they are not energy stores, but rather energy converters, they work similarly to batteries. They consist of anode, cathode and electrolyte membrane. Hydrogen is passed through the anode and oxygen through the cathode. The hydrogen is split into electrons and protons. As in a battery, the electrons go through the attached circuit and the protons go through the electrolyte membrane. At the cathode, the electrons, protons and the oxygen combine again to form water, which is the only byproduct of this reaction apart from heat. Fuel cells differ from batteries in that they need a constant flow of hydrogen and oxygen, which is why fuel cells are not considered energy stores by themselves [26].

G. Hydrogen storage

Hydrogen is often stored in tanks. In order to maximize the amount of hydrogen that can be stored in a given tank, it is often compressed or liquefied. Pressures can reach up to 700 bar and in order to liquefy hydrogen it has to be cooled to -253°C. Compressing and liquefying hydrogen to store it adds costs to hydrogen handling. Liquefying for example, can use up to 30 % of the energy contained in the liquefied hydrogen [27].

H. Hydrogen Electrolyzer

Hydrogen electrolyzers use electricity to split water into hydrogen and oxygen. They can therefore be used to create hydrogen to power fuel cells. Electrolyzers work similarly to fuel cells. They also consist of an anode, cathode and electrolyte. At the anode, water is split into oxygen and positively charged hydrogen ions. The electrons from this reaction flow through an external circuit, which powers the electrolyzer and the hydrogen ions travel through the electrolyte to the cathode. At the cathode, electrons from the external circuit and the positively charged hydrogen ions form hydrogen gas, which can then be extracted. Electrolyzers can be used to store energy in conjunction with hydrogen storage, if energy created by, e.g., a wind turbine is not currently needed [28].

IV. Optimization problems

The goal of optimization is generally to tune parameters of a system in a way, that makes the resulting system optimal. In order to know what is optimal, an optimization goal needs to be defined, which measures the performance of a certain set of parameters. Formally, an optimization problem can be described as [29]:

$$Minimize/Maximize: \quad F(x) \tag{1}$$

$$subject\ to: \quad g_j(x) \leq 0;\ j = 1, 2, ..., m\ , \tag{2}$$

where $F(x)$ is the target function representing the optimization goal and x is the parameter vector. The problem might be subject to a total number of m constraints $g_j(x)$ that limit the solution space.

Commonly used optimization goals and techniques in HRES optimization will be explained later in this paper.

A. Multi-Objective optimization

In order to incorporate multiple optimization goals into the optimization process, multi-objective optimization is often used. It also finds usage in HRES optimization frequently [3]–[5] [30]–[32]. Multi-objective optimization allows for goals to be combined and aims to find solutions that are good compromises regarding different targets.

In general, a multi objective optimization problem can be described as [33]

$$Minimize/Maximize: \quad F_{mo}(x) = [F_1(x), F_2(x), ..., F_k(x)] \tag{3}$$

$$subject\ to: \quad g_j(x) \leq 0;\ j = 1, 2, ..., m\ , \tag{4}$$

where $F_{mo}(x)$ is the multi objective function to be optimized, containing k single objective functions $F_k(x)$ and $g_j(x)$ the constraints like above.

There are two main approaches to multi-objective optimization. One combines the different objective functions into a weighted sum [33]

$$F_{mo} = \sum_{i=1}^{k} w_i \cdot F_i\ , \tag{5}$$

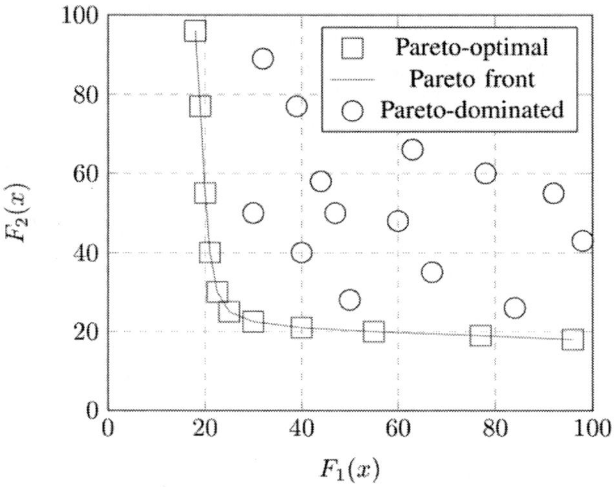

Figure 1. Example of Pareto front for 2 dimensional minimization problem

where w_i is the weight or importance of objective i and $F_i(x)$ is the ith objective function. In order to make this work, the different objective functions must either all be minimized or all be maximized. This can be achieved by multiplying objective functions by -1 to get the desired direction. The individual objective functions might also be normalized to allow the weight choice to directly represent an objectives importance without taking range of values into account [30]. For the other approach, each single objective function is evaluated separately. A set of solutions is retained, within which each solution is Pareto optimal. Pareto optimality is based on the hypothesis that solutions cannot be compared if one is better at one objective and the other at another. A solution is only better if it is better for at least one objective and at least equally good on all other objectives. In this case, the better solution Pareto-dominates the other solution. Formally a solution x Pareto-dominates a solution x' in a minimization problem if and only if

$$\forall i \in 1, ..., k : F_i(x) \leq F_i(x'),$$
$$and\ \exists j \in 1, ..., N : F_i(x) < F_i(x') \tag{6}$$

Figure 1 shows an example of Pareto-optimal and Pareto-dominated solutions and the Pareto front for a 2 dimensional minimization problem.

The set of solutions contains only solutions that are Pareto-optimal, meaning that all solutions within the set are not comparable and they Pareto-dominate every other solution. This set is called Pareto-optimal set or Pareto front [34].

This approach does not result in one best solution, but a set of solutions, from which a human can choose a suitable solution for the underlying problem. It is also regularly used in the context of HRES optimization [3][32][35].

V. Optimization methods

A. Evolutionary Algorithms

Evolutionary Algorithms (EAs) are optimization algorithms inspired by evolution theory. They use a combination of recombination, mutation and selection operators to find good solutions in the search space. In general, EAs first create an initial set of solutions, called population. For each iteration of the algorithm, solutions undergo the aforementioned steps. First, individuals recombine, which combines parts of two or more individuals into one or more so called offspring solutions. Those are then mutated, which adds random changes. From the old population, called parents and the new offsprings, a new population is selected, which is the new generation of parents for the next algorithm iteration. Selection is done by comparing the fitness of individuals, which is measured with objective functions such as those described in the previous section. They also work well on multi-objective optimization problems [29].

EAs are used frequently to optimize HRESs [32][36]–[38].

B. Particle Swarm Optimization

Another common technique in HRES optimization is Particle Swarm Optimization (PSO) [5][30][35][39]. It is inspired by the behavior of biological swarms and was first introduced by Kennedy and Eberhart [40]. Similar to EAs, PSO uses multiple individuals to carry out the search. In this case the population is called swarm, and the individuals particles.

Each particle in the swarm has knowledge about it's personal best and the global best solution that was previously found, as well as a velocity of it's movement through search space. This velocity depends on the distance to the current personal and global optima. It is lower, the closer the current solution is to the optima and higher, if it is further away. This encourages exploitation close to and exploration far from known optima.

In an iteration of the algorithm, the particle positions are updated by adding the velocity to their current position. Next, the fitness values of the particles are evaluated and the current personal and global optima are updated. Finally, the new velocities are calculated [29].

C. Other optimization methods

Less popular optimization methods used for HRES optimization include *Honey Bee Mating Optimization* [3], *Ant Colony Optimiztion* [41], *Harmony Search* [42], *Sampling Average Method*[31], *Simulated Annealing* [43] and *Tabu Search* [44]. This is not an exhaustive list, since in theory every optimization technique is usable in HRES optimization. Hybrid optimization methods that combine those algorithms are also investigated [45][46].

D. Software Solutions

Many software solutions exist for modeling, simulating and also optimizing HRES. Their main advantage is allowing optimization of HRES for users without algorithmic and programming skills. They also include modeling and simulation, which makes it possible to optimize and evaluate with one single program. Cuesta et al. [47] recently carried out a study of those software solutions.

The most popular HRES tool is *HOMER*. It allows optimization of systems regarding the Net Present Cost (NPC) [47]. Although this is the only possible optimization goal when using HOMER, it is frequently used for HRES optimization [4][48]–[50].

According to the study of Cuesta et al. other software capable of optimization are *DER-CAM*, *iHoga* and the open source *Calliope*. Calliope allows only optimization regarding Cost of Energy generation (COE), DER-CAM COE and CO_2 emissions, iHOGA NPC, CO_2 emissions, loss of load, human development index, and job creation. DER-CAM and iHoga can make use of multi objective optimization, iHOGA even allows for Pareto-optimization. From those tools only iHOGA is used somewhat regularly in HRES optimization [49][51].

VI. Simulation Methods

A very important part of HRES optimization is the simulation of solutions. In order to calculate the target functions of the optimization, the performance of solutions must be evaluated. Since it is usually not feasible to build the proposed HRES and measure it's performance in real life, simulation is often employed.

In order to be able to simulate a HRES, energy demand and generation need to be modeled. Demand is modeled by load profiles. The energy generation of photovoltaic systems and wind turbines mainly depends on the weather. Because of that, sun radiation and wind speed profiles are important to calculate the output of those components. Practically, all of the surveyed work relies on those profiles to simulate HRES performance.

One common way to simulate the system is using mathematical models of the HRES components to calculate their output [5][30][32][35]–[37]. For every simulation step, the generated energy of renewable sources is calculated with equations that depend on the sun radiation or wind speed, as well as efficiency of the device. This generation is then compared to the load. If there is a surplus of energy, it is stored in the respective storage device of the system. Since those processes are not 100 % efficient, equations are used to calculate the stored energy depending on input and device efficiency. For electrolyzers, the amount of created hydrogen is calculated here. If not enough energy to meet the load is generated by the renewable sources, the stored energy is usually used first. Depending on the storage device it is calculated how much the storage has to be drained to supply the necessary load. In case the renewable and stored energy is not enough to supply the load, fallback solutions like diesel generators and gas turbines have to be used. The fuel consumption needed to supply the load can then be calculated by mathematical models as well.

Many different equations exist for modeling HRES components, which will not be discussed in detail here, because of varying complexity and accuracy. The review by Bhandari et al. [52] provides a good overview.

HRES tools such as the previously mentioned HOMER, DER-CAM and iHOGA include simulations of the created systems [47]. This is a key aspect of what makes these tools popular since no own implementation of the mathematical models is required.

Failures of components or other disturbances could be included into the simulation e.g, by deactivating certain energy sources for some time in the simulation run. This would mimic real life failures in the system and allow for the investigation of the systems robustness and resilience. For resilience testing, further logic would need to be implemented into the simulation models that allows for the downgrading of functionality by e.g, cutting power to certain loads or by removing the connection to parts of the system in order to stabilize the system as a whole. Of the reviewed publications, none have incorporated system disturbances or the ability to downgrade into their simulations and thus did not challenge the robustness or resilience of the system.

VII. Optimization Goals

Optimization goals are used as target functions for optimizing HRESs. Different configurations can be compared by comparing those target functions. This section gives an overview of commonly used goals in HRES optimization and evaluates their incorporation of robustness and resilience properties.

A. Economic optimization goals

A common economic optimization goal for HRESs is the COE [4][5][36]. The COE describes how expensive the average annual energy creation of a system is per unit of energy and is often given in USD/kWh. Similarly, the Levelized Cost of Energy generation (LCOE) describes the average energy creation cost per unit over the entire project lifespan [30]. The LCOE can be calculated as [30][53]

$$LCOE = \frac{TPV}{E_L} CRF \quad [\text{USD/kWh}] \;, \tag{7}$$

where the Total Present Value (TPV) depends on the components of the system:

$$TPV = \sum_{d=1}^{k} C_d \quad [\text{USD}] \;, \tag{8}$$

where d is the device of k total devices and C_d are the costs associated with said device calculated as:

$$C_d = Init_d + C_{O\&M_d} \quad [\text{USD}] \;, \tag{9}$$

comprised of initialization costs $Init_d$ and operation and maintenance costs $C_{O\&M_d}$, which also include replacement costs if necessary.

In equation 7, E_L $[kWh]$ is the total load over the simulation period. Capital recovery factor (CRF) takes the interest rate into consideration and is calculated as:

$$CRF = \frac{i(1+i)^n}{(1+i)^n - 1} \;, \tag{10}$$

where i $[\%]$ is the nominal interest rate and n $[years]$ is the system life.

The Net Present Value (NPV) is the difference between the present value of cash inflow and the present value of cash outflow of a system over a period of time and therefore a measurement for the return of investment [54]. In the context of HOMER (see Section V-D), the NPV is called NPC [55] and is used under that name in several publications [4][31].

Using the notation from above, it can be calculated as

$$NPV = (E_L - TPV)\; CRF \quad [\text{USD}] \;. \tag{11}$$

The LCOE is the selling price needed to yield a NPV of 0 [56].

The aforementioned measurements take into account costs and earnings. Goals focused exclusively on the costs of a system are also used in the context of HRES optimization.

One such approach is the Annualized Cost of System (ACS), which annualizes all costs of the entire system [32] and can be described as

$$ACS = \sum_{d=1} C_{ad} \quad [\text{USD}] \;, \tag{12}$$

where C_{ad} are the annualized costs of a device that occur over the project's lifespan. A similar approach is the Life Cycle Cost (LCC). It sums all costs over the project but does not annualize them [37].

The Initial Capital Cost (ICC) measures how high the initial investment of a system is. This can be useful in situations with limited initial budget and calculated as [37]

$$ICC = \sum_{d=1} Init_d \quad [\text{USD}] \;. \tag{13}$$

None of the presented economic optimization goals measure the robustness or resilience of the system. Since they are concerned with the cost of the system or the generated energy, they rather work contradictory to the idea of a robust and resilient system. Enabling a system to be able to handle disturbances usually means adding redundancies, which in turn increases costs. Balancing those opposing objectives would be a key challenge when robustness and resilience should be implemented into HRES optimization.

B. Technical optimization goals

Technical optimization goals aim to formulate target functions that measure service security and reliability of the system. A widespread technical optimization goal is the Loss of Power Supply Probability (LPSP) [5][30][32][39][57]. Some publications refer to it as Loss of Load Probability (LLP) [31][36]. It measures the probability of the system being unable to supply enough power to satisfy the energy demand at any given time and can be calculated as [58]

$$LPSP = \frac{\sum_{t=1}^{T} E_{DE}(t)}{\sum_{t=1}^{T} E_L(t)} \quad [\%] \;, \tag{14}$$

where t is the current time step of T total time steps of the simulation, $E_{DE}(t)$ the energy deficit at time step t and $E_L(t)$ the total load at time step t.

Apart from the usage as a regular optimization goal, the LPSP is often used as a constraint. In that case, solution candidates must achieve a LPSP under a chosen constant to even be considered. This is done to ensure a certain level of reliability while optimizing other aspects of the system.

The LPSP could be a measure of the system's robustness and resilience, **if** the system is exposed to disturbances. If the system was exposed to threats, the LPSP would be improved if the system was robust enough to withstand the threats. If the HRES also had the ability to downgrade it's functionality, the LPSP would increase less than in the case of a total collapse, which would then be a measure of resilience. In the reviewed publications that use it, the system is not exposed to such events and therefore the LPSP only measures the reliability in regular operation.

The minimization of power losses aims to improve the efficiency of a HRES [3]. It can be described as

$$Loss_P = \sum_{t=1}^{T}\sum_{i=1}^{N_{br}}(R_i \cdot |I_i|^2 \cdot \delta t) \quad [\mathrm{W\,h}] \ , \tag{15}$$

where i is the current branch, N_{br} is the total number of branches, R_i is the resistance of branch i, I_i is the actual current of branch i and δt is the time step in the simulation.

While the minimization of power losses is an important goal for creating an effective system, it has no direct impact on the robustness and resilience of the system.

C. Environmental optimization goals

One of the most present goals in HRES optimization is the reduction of emissions. The direct emission of CO_2 from the combustion processes within a diesel generator or gas turbines over the course of the project or a year is often used as a target function [31][32]. It can be calculated as [35]

$$Emission = \sum_{d=1}^{D}\sum_{t=1}^{T} cons_d(t) \cdot EF_d \quad [\mathrm{kg}] \ , \tag{16}$$

where $cons_d(t)$ is the fuel consumption of device d at time t and EF_d is an emission factor that is specific to the device's and the fuel's characteristics. The emission factor usually ranges from 2.4 to 2.8 kg/l [59].

Other approaches try to incorporate all CO_2 emissions of a device over it's entire lifetime [30]. This includes emissions from harvesting the used materials, manufacturing, transporting, installing, operating and maintaining the device, as well as disposing it [60]. By dividing the emissions by the amount of generated energy, the Carbon Footprint off Energy (CFOE) can be calculated. It quantifies the emission of equivalent CO_2 mass per kWh of produced energy and can be described as [30][61]

$$CFOE = \frac{\epsilon_{sys}}{E_L} \quad [\mathrm{kg}CO2_{eq}/\mathrm{kWh}] \ , \tag{17}$$

where ϵ_{sys} are total the emissions of the entire system:

$$\epsilon_{sys} = \sum_{d=1}^{D} \epsilon_d \quad [\mathrm{kg}\,CO2_{eq}] \tag{18}$$

and ϵ_d the total emissions of device d, which can be broken down into

$$\begin{aligned}\epsilon_d = {} & \epsilon_{mat} + \epsilon_{man} + \epsilon_{trans} + \epsilon_{inst} \\ & + \epsilon_{o\&m} + \epsilon_{disp} \quad [\mathrm{kg}\,CO2_{eq}] \ ,\end{aligned} \tag{19}$$

which are the emissions for material gathering, manufacturing, transporting, installing, operating and maintaining and disposing the device.

The Renewable Energy Ratio (RER) is the ratio of energy created by renewable sources vs. conventional sources an is used as a environmental optimization goal [31]. It can be calculated as

$$RER = \frac{E_{ren}}{E_{conv}} \ , \tag{20}$$

where E_{ren} and E_{conv} are the amounts of energy created by renewable and conventional sources respectively.

A similar approach is the Renewables Factor (RF) [30] calculated as

$$RF = 1 - \frac{E_{conv}}{E_{ren}} \ . \tag{21}$$

None of the described goals measure robustness and resilience or directly impact those properties.

D. Socio-Political optimization goals

Recently, socio-political optimization goals are being scrutinized, since HRES impact communities in which they are installed beyond technical or environmental criteria, e.g., by creating jobs or shaping the landscape [47].

Eriksson et al. [30] have proposed a way to quantify the socio-political impact of a HRES to include it into the optimization process. The approach incorporates qualitative and quantitative factors to create an index-based measurement that represents the expected public satisfaction of a HRES. The used parameters are:

- **Aesthetics:** Acceptance of visual appearance, noise disturbance etc.
- **Employment:** Employment opportunities
- **Perceived hazard:** Potential hazard risk
- **Land requirement and acquisition:** Public resistance to land acquisition
- **Perceived local environmental impact:** Impact such as eco-system disturbances
- **Local ownership:** Ratio of local ownership in the proposed system
- **Local skills availability:** Availability of local workforce suitable for the project
- **Local resource availability:** Availability of local resources needed for the project
- **RF:** Penalty for reliance on non-renewable energy
- **Perceived service ability:** Level of improved service ability, such as improved availability of social electricity services

A score is assigned to each parameter ranging from 1 to 5 in order to rate a system or component. The scores are weighed to represent their importance on the given project, since every project has different priorities and different factors are important to the community. The weighed scores are then summed to give a single score named *Socio*. The perceived service ability parameter of the socio is impacted by a robust or resilient system, since the availability of the electrical services rises with robustness and resilience. Since no disturbances were used in it's original publication, that potential metric of robustness and resilience is yet unused.

VIII. Research Gap

Some work has been done on resilience of HRES without specifically targeting optimization. Kosai et al. [62] proposed a method to analyze system resilience regarding batteries. They measure resilience of a HRES by assessing how much of the batteries can fail for how long throughout a day without impacting self sustainability of the system. Using those two performance indices, the authors size batteries of HRES to have sufficient resilience at minimal cost. Approaches like these only mimic failures and downgrading of certain components and cannot sufficiently evaluate the entire systems vulnerability to attacks or disasters due to this.

Currently, robustness and resilience is not considered in HRES optimization, as we have explained in the previous sections. Some of the optimization goals like the LPSP (see Section VII-B) and the *Socio* (see Section VII-D) have the potential to measure robustness and resilience of the system. In the reviewed publications, they could not fulfill that potential, because no disturbances were incorporated while simulating system performance. Also, none of the systems had the ability to downgrade their functionality (see Section VI), which is the key part of a resilient system according to the definitions we showed. Recent cyber attacks, e.g., the blackout in Ukraine [10] and the increasing amounts of natural disasters like earthquakes, storms and floods, as well as disturbances due to system overloads [11] show that energy grids can be disrupted in unforseeable ways [12]. Because of this, it would make sense to include resilience and robustness against such events into HRES optimization by challenging the systems with these occurrences.

In summary, a research gap exists in scrutinizing HRES robustness and resilience by confronting them with disruptions. Also, no distinct measures of system resilience or robustness against such disruptions have been used in HRES optimization.

IX. Conclusion

This paper provided an overview of HRES optimization by explaining frequently used techniques to optimize and simulate HRES. Since no commonly agreed upon definition exists for resilience and robustness of cyber-physical systems, a selection of definitions from literature was presented and applied to the energy grid and in extension HRES in Section II. We explained common optimization goals and techniques in Sections V-VII, which are summarized in tables I and II and highlighted a lack of consideration of robustness and resilience against unexpected disruptions in literature.

TABLE I
SUMMARY OF OPTIMIZATION GOALS

Goal	Formula
	Economic optimization goals
LCOE	$LCOE = \frac{TPV}{E_L} CRF$ [USD/kWh]
NPV	$NPV = (E_L - TPV)\ CRF$ [USD]
ACS	$ACS = \sum_{d=1} C_{ad}$ [USD]
ICC	$ICC = \sum_{d=1} Init_d$ [USD]
	Technical optimization goals
LPSP	$LPSP = \frac{\sum_{t=1}^{T} E_{DE}(t)}{\sum_{t=1}^{T} E_L(t)}$ [%]
Power loss	$Loss_P = \sum_{t=1}^{T} \sum_{i=1}^{N_{br}} (R_i \cdot \lvert I_i \rvert^2 \cdot \delta t)$ [W h]
	Environmental optimization goals
CO_2 emission	$Emission = \sum_{d=1}^{D} \sum_{t=1}^{T} cons_d(t) \cdot EF_d$ [kg]
CFOE	$CFOE = \frac{\epsilon_{sys}}{E_L}$ [kg$CO2_{eq}$/kWh]
	Socio-Political optimization goals
Socio	Weighted score of multiple parameters

TABLE II
SUMMARY OF OPTIMIZATION AND SIMULATION METHODS

Optimization method	Possible goals	Simulation method
EA	All	Mathematical modeling
PSO	All	Mathematical modeling
HOMER	NPC	Internal simulation
Calliope	COE	Internal simulation
DER-CAM	COE CO_2 emission	Internal simulation
iHOGA	NPC CO_2 emission LLP	Internal simulation

In order to address this research gap, we will develop optimization goals that measure system reliability and robustness in future work. The main concept used for this will be Adversarial Resilience Learning (ARL) [13][63][64]. It allows two agents to compete on the same environment. In the context of ARL, those agents usually take the role of attacker and defender and one key showcase of the concept is the energy grid. Here, the attacker tries to destabilize the grid and the defender tries to keep the grid in a stable state. It is possible to use many different types of agents such as rule based, learning or random agents. The defender could be realized as a multi-agent system, as those have been frequently used in smart grid management applications [65]–[70]. Learning agents can be used to uncover vulnerabilities of the underlying environment by learning attack strategies in the attacker agent. This approach could be expanded to HRES optimization by using proposed HRES configurations as environments and analyzing how easy the attacker agent could disrupt it. Possible measures in terms of optimization goals could be the time needed to disrupt the system, the amount of successful disruptions over multiple experiments, the ability of the system to downgrade and recover from those attacks or the time needed to fully recover. This could be integrated nicely into an optimization loop with the python framework *palaestrai* [64][71]. Palaestrai allows for easy setups of ARL experiments from configuration files, which is well suited for changing environments (changing configuration of HRES).

With this new approach robustness and resilience could be considered in HRES optimization, which would improve reliability of those systems in the future.

ACKNOWLEDGMENT

This work has been funded by the German Federal Ministry for Economic Affairs and Energy (project number 01ME18002B). The authors would like to express their sincere and warm-hearted gratitude to their former colleague Norman Ihle, who played a key role in initiating the FRESH project.

REFERENCES

[1] UNFCCC. (2015) Paris agreement. [retrieved: Apr., 2021]. [Online]. Available: https://unfccc.int/sites/default/files/english_paris_agreement.pdf

[2] P. G. Arul, V. K. Ramachandaramurthy, and R. K. Rajkumar, "Control strategies for a hybrid renewable energy system: A review," *Renewable and Sustainable Energy Reviews*, vol. 42, pp. 597–608, 2015. [Online]. Available: http://dx.doi.org/10.1016/j.rser.2014.10.062

[3] T. Niknam, S. I. Taheri, J. Aghaei, S. Tabatabaei, and M. Nayeripour, "A modified honey bee mating optimization algorithm for multiobjective placement of renewable energy resources," *Applied Energy*, vol. 88, no. 12, pp. 4817–4830, 2011. [Online]. Available: http://dx.doi.org/10.1016/j.apenergy.2011.06.023

[4] R. Rajbongshi, D. Borgohain, and S. Mahapatra, "Optimization of PV-biomass-diesel and grid base hybrid energy systems for rural electrification by using HOMER," *Energy*, vol. 126, pp. 461–474, 2017. [Online]. Available: http://dx.doi.org/10.1016/j.energy.2017.03.056

[5] V. M. Sanchez *et al.*, "Techno-economical optimization based on swarm intelligence algorithm for a stand-alone wind-photovoltaic-hydrogen power system at south-east region of Mexico," *International Journal of Hydrogen Energy*, vol. 39, no. 29, pp. 16 646–16 655, 2014. [Online]. Available: http://dx.doi.org/10.1016/j.ijhydene.2014.06.034

[6] A. Helal, R. El-Mohr, and H. Eldosouki, "Optimal design of hybrid renewable energy system for electrification of a remote village in Egypt," *2nd International Conference on Communications Computing and Control Applications, CCCA 2012*, pp. 1–6, 2012.

[7] E. L. Eriksson and E. M. A. Gray, "Optimization and integration of hybrid renewable energy hydrogen fuel cell energy systems – A critical review," *Applied Energy*, vol. 202, pp. 348–364, 2017. [Online]. Available: http://dx.doi.org/10.1016/j.apenergy.2017.03.132

[8] F. R. Yu, P. Zhang, W. Xiao, and P. Choudhury, "Communication systems for grid integration of renewable energy resources," *IEEE Network*, vol. 25, no. 5, pp. 22–29, 2011.

[9] C.-C. Sun, C.-C. Liu, and J. Xie, "Cyber-physical system security of a power grid: State-of-the-art," *Electronics*, vol. 5, no. 3, 2016, [retrieved: Apr., 2021]. [Online]. Available: https://www.mdpi.com/2079-9292/5/3/40

[10] J. E. Sullivan and D. Kamensky, "How cyber-attacks in Ukraine show the vulnerability of the U.S. power grid," *Electricity Journal*, vol. 30, no. 3, pp. 30–35, 2017. [Online]. Available: http://dx.doi.org/10.1016/j.tej.2017.02.006

[11] ENTSO-E. (2021) System separation in the continental europe synchronous area on 8 january 2021 – 2nd update. [retrieved: Apr., 2021]. [Online]. Available: https://www.entsoe.eu/news/2021/01/26/system-separation-in-the-continental-europe-synchronous-area-on-8-january-2021-2nd-update/

[12] Y. Wang, C. Chen, J. Wang, and R. Baldick, "Research on Resilience of Power Systems under Natural Disasters - A Review," *IEEE Transactions on Power Systems*, vol. 31, no. 2, pp. 1604–1613, 2016.

[13] E. Veith, L. Fischer, M. Tröschel, and A. Nieße, "Analyzing Cyber-Physical Systems from the Perspective of Artificial Intelligence," in *Proceedings of the 2019 International Conference on Artificial Intelligence, Robotics and Control*. ACM, 2019, pp. 85–95.

[14] R. Arghandeh, A. Von Meier, L. Mehrmanesh, and L. Mili, "On the definition of cyber-physical resilience in power systems," *Renewable and Sustainable Energy Reviews*, vol. 58, pp. 1060–1069, 2016.

[15] Critical infrastructure security and resilience. [retrieved: Apr., 2021]. [Online]. Available: https://obamawhitehouse.archives.gov/the-press-office/2013/02/12/presidential-policy-directive-critical-infrastructure-security-and-resil

[16] ENTSO-E. (2018) Synchronous area framework (safa) policy 1: Load-frequency control and reserves. [retrieved: Apr., 2021]. [Online]. Available: https://eepublicdownloads.entsoe.eu/clean-documents/Publications/SOC/safa/1_-_Policy_on_Load-Frequency_Control_and_Reserves.pdf

[17] K. Shivarama Krishna and K. Sathish Kumar, "A review on hybrid renewable energy systems," *Renewable and Sustainable Energy Reviews*, vol. 52, pp. 907–916, 2015. [Online]. Available: http://dx.doi.org/10.1016/j.rser.2015.07.187

[18] A. B. Awan, "Performance analysis and optimization of a hybrid renewable energy system for sustainable NEOM city in Saudi Arabia," *Journal of Renewable and Sustainable Energy*, vol. 11, no. 2, pp. 025 905–1–17, 2019.

[19] L. M. Halabi and S. Mekhilef, "Flexible hybrid renewable energy system design for a typical remote village located in tropical climate," *Journal of Cleaner Production*, vol. 177, pp. 908–924, 2018. [Online]. Available: https://doi.org/10.1016/j.jclepro.2017.12.248

[20] M. S. Islam, "A techno-economic feasibility analysis of hybrid renewable energy supply options for a grid-connected large office building in southeastern part of France," *Sustainable Cities and Society*, vol. 38, pp. 492–508, 2018. [Online]. Available: http://dx.doi.org/10.1016/j.scs.2018.01.022

[21] A. Fahrenbruch and R. Bube, *Fundamentals Of Solar Cells: Photovoltaic Solar Energy Conversion*. Elsevier Science, 2012. [Online]. Available: https://books.google.de/books?id=eh7H-9cz-IoC

[22] D. Spera, *Wind Turbine Technology: Fundamental Concepts of Wind Turbine Engineering*. ASME Press, 2009. [Online]. Available: https://books.google.de/books?id=hE1pPgAACAAJ

[23] H. Kiehne, *Battery Technology Handbook*, ser. Electrical and computer engineering. CRC Press, 2003. [Online]. Available: https://books.google.de/books?id=1HSsx9fPAKkC

[24] L. Mahon, *Diesel Generator Handbook*. Elsevier Science, 1992. [Online]. Available: https://books.google.de/books?id=S-RSAAAAMAAJ

[25] M. Boyce, *Gas Turbine Engineering Handbook*, ser. Chemical, Petrochemical & Process. Gulf Professional Pub., 2002. [Online]. Available: https://books.google.de/books?id=nEc2OxxT_uMC

[26] S. Revankar and P. Majumdar, *Fuel Cells: Principles, Design, and Analysis*, ser. Mechanical and Aerospace Engineering Series. CRC Press, 2016. [Online]. Available: https://books.google.de/books?id=c47SBQAAQBAJ

[27] R. Gupta, A. Basile, and T. Veziroglu, *Compendium of Hydrogen Energy: Hydrogen Storage, Distribution and Infrastructure*, ser. Woodhead Publishing Series in Energy. Elsevier Science, 2016. [Online]. Available: https://books.google.de/books?id=U24ZBgAAQBAJ

[28] M. Nehrir and C. Wang, *Modeling and Control of Fuel Cells: Distributed Generation Applications*, ser. IEEE Press Series on Power Engineering. Wiley, 2009. [Online]. Available: https://books.google.de/books?id=keg9JLtETh8C

[29] O. Kramer, *A Brief Introduction to Continuous Evolutionary Optimization*, ser. SpringerBriefs in Applied Sciences and Technology. Springer International Publishing, 2013. [Online]. Available: https://books.google.de/books?id=22vABAAAQBAJ

[30] E. L. Eriksson and E. M. A. Gray, "Optimization of renewable hybrid energy systems – A multi-objective approach," *Renewable Energy*, vol. 133, pp. 971–999, 2019. [Online]. Available: https://doi.org/10.1016/j.renene.2018.10.053

[31] M. Sharafi and T. Y. ElMekkawy, "Stochastic optimization of hybrid renewable energy systems using sampling average method," *Renewable and Sustainable Energy Reviews*, vol. 52, pp. 1668–1679, 2015.

[32] Z. Shi, R. Wang, and T. Zhang, "Multi-objective optimal design of hybrid renewable energy systems using preference-inspired coevolutionary approach," *Solar Energy*, vol. 118, pp. 96–106, 2015. [Online]. Available: http://dx.doi.org/10.1016/j.solener.2015.03.052

[33] R. T. Marler and J. S. Arora, "The weighted sum method for multi-objective optimization: New insights," *Structural and Multidisciplinary Optimization*, vol. 41, no. 6, pp. 853–862, 2010.

[34] P. Ngatchou, A. Zarei, and M. A. El-Sharkawi, "Pareto multi objective optimization," *Proceedings of the 13th International Conference on Intelligent Systems Application to Power Systems, ISAP'05*, vol. 2005, pp. 84–91, 2005.

[35] M. Sharafi and T. Y. ELMekkawy, "Multi-objective optimal design of hybrid renewable energy systems using PSO-simulation based approach," *Renewable Energy*, vol. 68, pp. 67–79, 2014.

[36] M. S. Ismail, M. Moghavvemi, and T. M. Mahlia, "Genetic algorithm based optimization on modeling and design of hybrid renewable energy systems," *Energy Conversion and Management*, vol. 85, pp. 120–130, 2014. [Online]. Available: http://dx.doi.org/10.1016/j.enconman.2014.05.064

[37] K. D. Mercado, J. Jimémez, and C. G. Quintero M., "Hybrid Renewable Energy System based on Intelligent Optimization Techniques," *International Conference on Renewable Energy Research and Applications*, vol. 5, pp. 661–666, 2016.

[38] M. Ming, R. Wang, Y. Zha, and T. Zhang, "Multi-objective optimization of hybrid renewable energy system using an enhanced multi-objective evolutionary algorithm," *Energies*, vol. 10, no. 5, pp. 5–9, 2017.

[39] A. Kashefi Kaviani, G. H. Riahy, and S. M. Kouhsari, "Optimal design of a reliable hydrogen-based stand-alone wind/PV generating system, considering component outages," *Renewable Energy*, vol. 34, no. 11, pp. 2380–2390, 2009. [Online]. Available: http://dx.doi.org/10.1016/j.renene.2009.03.020

[40] J. Kennedy and R. Eberhart, "Particle swarm optimization," in *Proceedings of ICNN'95-International Conference on Neural Networks*, vol. 4. IEEE, 1995, pp. 1942–1948.

[41] A. Fetanat and E. Khorasaninejad, "Size optimization for hybrid photovoltaic-wind energy system using ant colony optimization for continuous domains based integer programming," *Applied Soft Computing Journal*, vol. 31, pp. 196–209, 2015. [Online]. Available: http://dx.doi.org/10.1016/j.asoc.2015.02.047

[42] A. Chauhan and R. P. Saini, "Discrete harmony search based size optimization of Integrated Renewable Energy System for remote rural areas of Uttarakhand state in India," *Renewable Energy*, vol. 94, pp. 587–604, 2016. [Online]. Available: http://dx.doi.org/10.1016/j.renene.2016.03.079

[43] W. Zhang, A. Maleki, M. A. Rosen, and J. Liu, "Optimization with a simulated annealing algorithm of a hybrid system for renewable energy including battery and hydrogen storage," *Energy*, vol. 163, pp. 191–207, 2018.

[44] Y. A. Katsigiannis and P. S. Georgilakis, "Optimal sizing of small isolated hybrid power systems using tabu search," *Journal of Optoelectronics and Advanced Materials*, vol. 10, no. 5, pp. 1241–1245, 2008.

[45] W. Zhang, A. Maleki, M. A. Rosen, and J. Liu, "Sizing a stand-alone solar-wind-hydrogen energy system using weather forecasting and a hybrid search optimization algorithm," *Energy Conversion and Management*, vol. 180, no. August 2018, pp. 609–621, 2019. [Online]. Available: https://doi.org/10.1016/j.enconman.2018.08.102

[46] Y. A. Katsigiannis, P. S. Georgilakis, and E. S. Karapidakis, "Hybrid simulated annealing-tabu search method for optimal sizing of autonomous power systems with renewables," *IEEE Transactions on Sustainable Energy*, vol. 3, no. 3, pp. 330–338, 2012.

[47] M. A. Cuesta, T. Castillo-Calzadilla, and C. E. Borges, "A critical analysis on hybrid renewable energy modeling tools: An emerging opportunity to include social indicators to optimise systems in small communities," *Renewable and Sustainable Energy Reviews*, vol. 122, no. June 2019, p. 109691, 2020. [Online]. Available: https://doi.org/10.1016/j.rser.2019.109691

[48] S. Bahramara, M. P. Moghaddam, and M. R. Haghifam, "Optimal planning of hybrid renewable energy systems using HOMER: A review," *Renewable and Sustainable Energy Reviews*, vol. 62, pp. 609–620, 2016. [Online]. Available: http://dx.doi.org/10.1016/j.rser.2016.05.039

[49] N. Saiprasad, A. Kalam, and A. Zayegh, "Comparative Study of Optimization of HRES using HOMER and iHOGA Software," *Journal of Scientific and Industrial Research (JSIR)*, vol. 77, no. 12, pp. 677–683, 2018.

[50] S. Vendoti, M. Muralidhar, and R. Kiranmayi, "HOMER Based Optimization of Solar-Wind-Diesel Hybrid System for Electrification in a Rural Village," *2018 International Conference on Computer Communication and Informatics, ICCCI 2018*, pp. 1–6, 2018.

[51] P. Ganguly, A. Kalam, and A. Zayegh, "Design an optimum standalone hybrid renewable energy system for a small town at Portland, Victoria using iHOGA," *2017 Australasian Universities Power Engineering Conference, AUPEC 2017*, vol. 2017-November, pp. 1–6, 2018.

[52] B. Bhandari, S. R. Poudel, K. T. Lee, and S. H. Ahn, "Mathematical modeling of hybrid renewable energy system: A review on small hydro-solar-wind power generation," *International Journal of Precision Engineering and Manufacturing - Green Technology*, vol. 1, no. 2, pp. 157–173, 2014.

[53] B. Sørensen, *Solar Energy Storage*. Elsevier Science, 2015. [Online]. Available: https://books.google.de/books?id=Qs-cBAAAQBAJ

[54] A. Pechmann, I. Schöler, and S. Ernst, "Possibilities for CO2-neutral manufacturing with attractive energy costs," *Journal of Cleaner Production*, vol. 138, pp. 287–297, 2016.

[55] HOMER Energy. (2020) Homer pro documentation - net present cost. [retrieved: Apr., 2021]. [Online]. Available: https://www.homerenergy.com/products/pro/docs/latest/net_present_cost.html

[56] I. Pawel, "The cost of storage - How to calculate the levelized cost of stored energy (LCOE) and applications to renewable energy generation," *Energy Procedia*, vol. 46, pp. 68–77, 2014. [Online]. Available: http://dx.doi.org/10.1016/j.egypro.2014.01.159

[57] H. Yang, L. Lu, and W. Zhou, "A novel optimization sizing model for hybrid solar-wind power generation system," *Solar Energy*, vol. 81, no. 1, pp. 76–84, 2007.

[58] D. Abbes, A. Martinez, and G. Champenois, "Life cycle cost, embodied energy and loss of power supply probability for the optimal design of hybrid power systems," *Mathematics and Computers in Simulation*, vol. 98, pp. 46–62, 2014. [Online]. Available: http://dx.doi.org/10.1016/j.matcom.2013.05.004

[59] C. Borgnakke and R. Sonntag, *Fundamentals of Thermodynamics, 9th Edition*. Wiley, 2016. [Online]. Available: https://books.google.de/books?id=zcOPDQAAQBAJ

[60] R. García-Valverde, C. Miguel, R. Martínez-Béjar, and A. Urbina, "Life cycle assessment study of a 4.2 kWp stand-alone photovoltaic system," *Solar Energy*, vol. 83, no. 9, pp. 1434–1445, 2009.

[61] M. Bortolini, M. Gamberi, A. Graziani, and F. Pilati, "Economic and environmental bi-objective design of an off-grid photovoltaic-battery-diesel generator hybrid energy system," *Energy Conversion and Management*, vol. 106, pp. 1024–1038, 2015. [Online]. Available: http://dx.doi.org/10.1016/j.enconman.2015.10.051

[62] S. Kosai and J. Cravioto, "Resilience of standalone hybrid renewable energy systems: The role of storage capacity," *Energy*, vol. 196, p. 117133, 2020. [Online]. Available: https://doi.org/10.1016/j.energy.2020.117133

[63] L. Fischer, J. Memmen, E. M. S. P. Veith, and M. Tröschel, "Adversarial resilience learning - towards systemic vulnerability analysis for large and complex systems," *CoRR*, vol. abs/1811.06447, pp. 24–32, 2018. [Online]. Available: http://arxiv.org/abs/1811.06447

[64] E. M. Veith, N. Wenninghoff, and E. Frost, "The Adversarial Resilience Learning Architecture for AI-based Modelling, Exploration, and Operation of Complex Cyber-Physical Systems," 2020, [retrieved: Apr., 2021]. [Online]. Available: http://arxiv.org/abs/2005.13601

[65] S. Holly *et al.*, "Flexibility management and provision of balancing services with battery-electric automated guided vehicles in the hamburg container terminal altenwerder," *Energy Informatics*, vol. 3, no. 1, pp. 1–20, 2020.

[66] L. Hernández *et al.*, "A multi-agent system architecture for smart grid management and forecasting of energy demand in virtual power plants," *IEEE Communications Magazine*, vol. 51, no. 1, pp. 106–113, 2013.

[67] A. Nieße *et al.*, "Market-based self-organized provision of active power and ancillary services: An agent-based approach for smart distribution grids," in *2012 Complexity in Engineering (COMPENG). Proceedings*. IEEE, 2012, pp. 1–5.

[68] R. Schwerdfeger and D. Westermann, "Approach for ancillary service provision by decentralized energy devices with the focus on load frequency control," 2014, [retrieved: Apr.,

2021]. [Online]. Available: http://b-dig.iie.org.mx/BibDig2/P15-0324/PDFfiles/PESGM2015-000820.PDF

[69] H.-M. Kim, Y. Lim, and T. Kinoshita, "An intelligent multiagent system for autonomous microgrid operation," *Energies*, vol. 5, no. 9, pp. 3347–3362, 2012.

[70] W. Li, T. Logenthiran, V.-T. Phan, and W. L. Woo, "Intelligent multi-agent system for power grid communication," in *2016 IEEE Region 10 Conference (TENCON)*. IEEE, 2016, pp. 3386–3389.

[71] GitLab. (2020) palestrai GitLab Page. [retrieved: Apr., 2021]. [Online]. Available: https://gitlab.com/arl2/palaestrai

Non-Intrusive Load Monitoring of Single and Aggregated Profiles with a Hidden Markov Model

Nadège Miquey
Ecole centrale de Lyon
Lyon, France
email: nadege.miquey@ecoco2.com

Etta Grover-Silva
Eco CO2
Nanterre, France
email: etta.grover-silva@ecoco2.com

***Abstract*—Awareness raising programs to encourage energy efficient behaviour are important in the context of the current energy transition. Sensors and connected devices allowing for data collection are easily available providing an opportunity to collect electric consumption data from individual appliances. The effective use of these data sources is necessary to optimize an energy efficiency coaching program. This paper presents a methodology for non-intrusive load monitoring analysis on individual smart plugs to identify an unknown appliance and disaggregated an aggregated load profile, such as a power strip. The automatic detection of a connected appliance allows for appliance usage suggestions to be provided quickly without the need of an end-user input. The disaggregation of an aggregated curve such as a power strip allows for the optimization of the number of smart plugs required to represent a significant proportion of the total energy use of the household. The down sampling of high resolution data was also performed to observe the performance of the methodology on lower resolution data. The single appliance identification models all had very high accuracy (between 94 - 100 %). The disaggregation of an aggregated profile in the kitchen use case also had high accuracy for data with a resolution of less than one minute (95 - 99 %). The disaggregation of an aggregated profile for a multi-media use case had a lower performance when more than two appliances were considered (55 - 85 %).**

Keywords*—*load monitoring, machine learning, disaggregation, energy efficiency.

I. Introduction

The energy transition has become an urgent topic in the context of the accelerating climate change. This transition implies the significant reduction of resource usage in order to accomplish ambitious goals set forth by world leaders including the European Union (EU). The building sector and more precisely the residential building sector is a large contributor to the energy consumption of a country. In the European Union, households represent one-fourth of its total final energy consumption [1]. In recent years, emissions as a result of building energy consumption have increased reaching an all-time high in 2018 [2].

The energy consumption of a household is dependent on the electric appliances present and the use of these appliances defined by individual appliance usage behaviour. The second factor can have a huge impact on the overall energy use. Inefficient use of building systems and appliances is called the energy efficiency gap. This gap can represent up to 100% of excessive energy use in comparison to the design requirements [3]. The EU has put in place strict regulations about new building construction to improve energy efficient buildings, however few regulations exist addressing the behavioural use of buildings after construction.

Energy efficient behaviour is highly specific to individual households. Therefore, effective awareness raising programs are difficult to implement on a large scale without integrating household specific analysis and advice. Several studies have shown that appliance specific energy use combined with real-time feedback results in the highest energy savings for multiple awareness raising programs [4].

With recent massive deployment of smart meters in multiple countries, individual household electric load data is more easily accessible. Multiple awareness raising programs have been deployed by the company Eco CO2 in the context of a public tender in France put forth by ADEME related to a funding mechanism called *Investissement d'avenir*. The programs deployed include TBH Alliance [5], Picowatty [6] and SEIZE [7]. These programs focus on encouraging energy efficient behaviour by providing useful and user specific information about the environment and the direct impact of their actions.

Recently, individual electric appliance load plugs have been integrated in Eco CO2 data collection platform. These smart plugs allow for the precise measurement of individual appliance load profiles or of a power strip to provide appliance specific advice to individual users. However, the cost of individual smart plugs does not allow for the individual load monitoring of all appliances within a household. In order to maximize the impact of a single smart plug, Non-Intrusive Load Monitoring (NILM) techniques [8][9] can be applied to decompose a combined load curve of up to 6 appliances connected to a power strip.

Multiple algorithms have been implemented in the NILM domain for the classification of appliance profiles including K-Nearest Neighbors (KNN) [10][11], K-means [12], Decision Tree, Random Forest, Recurrent Neural Network (RNN) [13] and Hidden Markov Model (HMM) [14]. The selection of the appropriate method is based on the data available for model training, the data resolution and the computational requirements.

In [15] individual appliance load data [16] with a one second resolution are modeled with a Random Forest, LogitBoost, Bagging, Decision Tree, Naive Bayes and Support Vector

Machine algorithms. The results were compared for the classification of 33 devices. Accuracy ranged from 90-96% with the Random Forest algorithm performing the best. However, these algorithms were not demonstrated to be capable of analyzing aggregated load profiles as well.

The KNN, K-means and Decision Tree algorithms are well adapted for single appliance identification. In [11] 10 second resolution data of active and reactive power is used for the correct classification of 8 individual appliances. One of these algorithms, the Feed-forward Neural Network, was also applied to aggregated curves to identify individual appliances within an aggregated profile. The aggregated model used three appliances with a 90-98% accuracy depending on the combined states.

A review of event based and non-event based NILM schemes is presented in [14] where two main techniques are discussed: HMM and Convolutional Neural Networks (CNN). The CNN and more generally Deep Neural Networks (DNN) approach was deemed more computationally intensive for the training period and relatively lower accuracy (below 90%) in comparison to HMM techniques. The HMM is classified as a non-event based NILM scheme and is preferential due to the effective application on individual and aggregated curves, the simplicity of the model training and the application on low resolution data.

In [9] advantages of HMM in comparison to other NILM techniques are described as being effective when labeled data is available and on low frequency timeseries data. These models can also have a low computational requirement depending on the model structure. This type of model has been proven to be effective on the identification of single appliances as well as aggregated profiles. It has been noted that this model is highly effective for multi-state appliances with distinct power levels but not ideal for multi-state variable power consumption appliances. The HMM is more effective in comparison to a Markov Model in a disaggregation context due to the fact that the results are not a simple sequence of states and the transition probability. With the HMM, it is possible to study observable power values associated with hidden states of appliances, the probability that each hidden state is a realistic combination of observed power states for each appliance power value state and the transition state probability.

In this paper, a HMM methodology is implemented due to the effectiveness of the model structure to classify profiles for an individual appliance as well as an aggregated load profile. It has been proven to be effective on low resolution data which is important for commercial application of the model and the required data storage necessary for effective classification. The training period is relatively simple and has a low computational requirement. The implementation of this model is innovative to help in providing targeted energy saving advice per appliance. Based on initial appliance classification and individual appliance consumption analysis, awareness raising programs and impact can be more closely studied.

The following sections will detail the simple but robust model developed for the automatic identification of individual appliances that is also applicable to a power strip to identify individual appliances in an aggregated load profile. The methodology is presented in section II followed by a realistic case study in section III. Results are discussed in IV which includes a performance study of the developed model on varying data resolutions.

II. Methodology

In this section, the Hidden Markov Model and pre-processing steps of constructing the model features will be explained. The pre-processing and feature extraction of the timeseries data is composed of three main steps: clustering of timeseries values, definition of buckets, feature quantification of buckets. These three steps allow for the effective quantification of the timeseries data to then develop individual appliance Hidden Markov Models for individual appliance recognition. These individual appliance models are then combined to produce an effective state identification model to separate individual appliance timeseries signals out of an aggregated timeseries signal.

A. Data pre-processing

1) Clustering on power consumption values: For each appliance, a K-means clustering method is used to identify groups of power consumption values and define the number of states characteristic of each appliance. The clustering method is applied to timeseries data for a duration of 2 to 10 days specified in Table I for each single appliance. The duration of training timeseries is chosen to get a representative number of active periods for each appliance. For appliances functioning more than 6 hours a day such as refrigerators, screen or Internet router, 3 days allow to identify the functioning range of values. For appliances functioning occasionally, up to 10 days of data are necessary to identify the range of values of functioning periods.

TABLE I
K-MEANS CLUSTERING, SAMPLING RATE = 5 SECONDS

Appliance type	Duration of training timeseries (days)	Number of active periods
Hot-water boiler	10	21
Refrigerator	3	94
Coffee-machine	10	29
Washing-machine	10	10
Screen	3	8
Internet router	3	NA[a]
Laptop charger	6	8
Television	5	12

[a]Not Applicable, Internet router is an always on appliance.

A silhouette score is computed to determine the ideal number of clusters for each appliance. This method is used as opposed to a possible K-means clustering methodology in order to reduce computational time and prioritize simplicity of the calculation. The average power value associated with each cluster is also calculated.

2) Interpolation and bucketing of aggregated profile: Load profile data is assigned labels based on assigned clusters from the K-means clustering method. This allows for the reconstruction of the timeseries data with labeled time periods associated with each K-means cluster. An average value is then assigned to each cluster and used to remove all variation within the period designated as the same cluster label. An interpolation method is used to fill in missing data and reconstruct a continuous timeseries profile. This new constructed load profile is then separated into buckets based on the magnitudes evaluated in the K-means clustering method.

3) Test and training data processing: Characteristics of the testing data set can differ from the characteristics of the training data set. To avoid this, load profiles are constructed to obtain test load profiles closer to training load profiles and improve models performances. To construct the load profiles, buckets are created around each cluster's value identified on training load profiles. For the appliance i, with n_i clusters of mean values $m_1, m_2, ..., m_n$, the created buckets are $m_1 - x\%, m_1 + x\%, m_2 - x\%, ..., m_n + x\%$. Models performances are evaluated and compared on raw test data and on pre-processed load profiles.

B. Hidden Markov Model

The calculated features are then used to construct individual Hidden Markov Models for each individual appliance. A multi-appliance model is then created by combining all possible states of the individual appliance models.

1) Single appliance model: The ideal number of clusters computed with the K-means clustering method is used to set the number of hidden states for each single Hidden Markov Model. Parameters of each single appliance HMM are estimated with an Expectation–Maximization (EM) algorithm.

2) Multi-appliance model: Single appliance models are then combined to obtain Hidden Markov Model for multiple appliances. Single appliance models states are combined to form N distinct combinations of states of the combined model.

$$N = \prod_{i=0}^{k} n_i \tag{1}$$

where $k \in \mathbb{N}^*$ is the number of appliances combined in the model, $n_i \in \mathbb{N}^*$ is the number of states of the ith appliance.

Transition matrices of single appliance models are combined using Kronecker product defined by (2):

$$\mathbf{A} \otimes \mathbf{B} = \begin{pmatrix} a_{1,1}\mathbf{B} & \cdots & a_{1,n}\mathbf{B} \\ a_{2,1}\mathbf{B} & \cdots & a_{2,n}\mathbf{B} \\ \vdots & \ddots & \vdots \\ a_{m,1}\mathbf{B} & \cdots & a_{m,n}\mathbf{B} \end{pmatrix} \tag{2}$$

Viterbi algorithm is used to decode sequences and determine the most probable sequence of hidden states (which appliance is in which states) corresponding to the observable sequence of power consumption traces.

3) Splitting data into a training set and a test set: For each appliance class, the load profiles of a specific appliance are used to train the HMM (different refrigerators, washing-machines etc). Feature extraction using bucketing allows to identify specific characteristics for each appliance in the same class (specific load profile patterns, maximum power, average power etc). Testing of the developed model is then performed on load profiles that were not included in the training set.

C. Single model evaluation metrics

Single HMM are evaluated on 15 days of data using the following metrics:

- the accuracy: the proportion of correct On and Off states prediction
- the precision defined in (3): the proportion of correct On states prediction among all On states predicted. TP is the number of True Positive elements, FN is the number of False Negative elements

$$precision = \frac{TP}{TP + FN} \tag{3}$$

- the f1-score defined in (4), where precision is defined in (3) and recall defined in (5)

$$F1 = 2 * \frac{precision * recall}{precision + recall} \tag{4}$$

$$recall = \frac{TP}{TP + FN} \tag{5}$$

D. Combined model evaluation metrics

To evaluate performances of states prediction, two different metrics are used. A micro-averaged $accuracy$ score defined in (6) is computed. This score measures the fraction of correct states prediction among all possible combination of states of the combined model. When power consumption values are below a certain threshold, they are bucketed to zero consumption values. States where all appliances are Off are thus the easiest ones to detect and these states concern a large part of each recording. To get a more representative accuracy score, the micro-averaged $accuracy_{on}$ score defined in (7) does not take into account the state where all appliances are Off.

$$micro\_accuracy = \frac{\sum_{n=0}^{N} TP_n}{\sum_{n=0}^{N} TP_n + \sum_{n=0}^{N} FP_n} \tag{6}$$

$\forall$n $\in [0, N]$ such that $n \neq$ *off state* :

$$micro\_accuracy_{on} = \frac{\sum_{n=0}^{N} TP_n}{\sum_{n=0}^{N} TP_n + \sum_{n=0}^{N} FP_n} \tag{7}$$

Where:

- N is the number of states defined in 1.
- TP_n refer to True Positive elements of states n: predicted states are n and correspond to ground truth.
- FP_n refer to False Positive element of states n: predicted states are n whereas ground truth is not n.

III. Case study

The primary objective of this case study is to show the effectiveness of the defined methodology to automatically classifying individual appliances as well as combined appliances load profiles. A combined appliance load profile could be a smart plug monitoring a power strip where up to six appliances are connected. The data used for testing and validation was collected through an internal experimental study performed with employees of Eco CO2. The recorded power consumption data was collected for 64 smart plugs. More than 3800 days of data have been collected during the experimental study. The appliances used in this case study include refrigerators, coffee-machines, hot-water boilers, microwave-ovens, Internet router, screens, computers, televisions, washing-machines. Two realistic power strip configurations have been defined to simulate a realistic application of a power strip found in a typical residential household:

- Aggregated load profile of four kitchen appliances: a hot-water boiler, a refrigerator, a coffee-machine, a washing-machine.
- Aggregated load profile of four multimedia appliances: a screen, an Internet router, a laptop charger and a television.

Three models are then computed for each category: a combined HMM model of two single appliance models, a combined HMM model of three single appliance models and a combined HMM model of four single appliance models.

IV. Results

In this section, the results of the K-means clustering method applied on eight single appliances will be detailed. State prediction performances of the HMM defined on single appliances and aggregated profiles will be evaluated. The aggregated profiles represent two realistic configurations of possible power strips in a kitchen or office scenario presented in section IV. A comparison between results on kitchen appliances and on multimedia appliances is done based on the specificity of each appliance type. The impact of data resolution degradation on combined appliance HMM states predictions will be studied.

A. Model prediction of on and off states evaluation

Table II presents each cluster number and cluster centroide for each appliance computed with a K-means method. A majority of appliances analyzed could be coerced into a maximum two states. The one exception is the refrigerator category due to a high peak as a result of the electrical inertia of the compressor, therefore creating three distinct states.

Appliances with no zero state can be interpreted as loads that contribute to the standby consumption. Therefore, the appliance is never completely off and consumes electricity constantly even if the consumption is very low.

B. State prediction appliance models

1) State prediction of single appliance models: Each model is evaluated using the evaluation metrics defined in Section II-C to determine the accuracy of state prediction. Evaluation metrics are calculated for 15 periods of 24-hours. Each test sample is the addition of 24-hours recordings of single appliances present in the tested model.

TABLE II
K-MEANS RESULTS, SAMPLING RATE = 1 SECONDS

Appliance category	Appliance type	Number of clusters	Clusters centroids (W)
Kitchen appliances	Hot-water boiler	2	[0.3839, 2468]
	Refrigerator	3	[0.1490, 117.7, 1269]
	Coffee-machine	2	[1.417, 1576]
	Washing-machine	2	[2.813, 2438]
Multimedia appliances	Screen	2	[1.0, 29.06]
	Internet router	2	[0.0, 8.122]
	Laptop charger	2	[13.04, 0.015]
	Television	2	[0.0, 129.4]

TABLE III
SINGLE APPLIANCE HMM RESULTS ON RAW DATA FOR KITCHEN APPLIANCES, SAMPLING RATE = 1 SECOND

Appliance category	Appliance type	Accuracy (%)	Precision	f1-score
Kitchen appliances	Hot-water boiler	99.9	0.82	0.90
	Refrigerator	99.7	NA[a]	NA[a]
	Coffee-machine	99.6	0.99	0.99
	Washing-machine	94.3	0.75	0.84

[a]Not Applicable, refrigerators are three-state appliances.

Table III shows the accuracy results of states prediction for each single appliance HMM when raw data are processed. Low accuracy results are observed for the washing-machine appliance due to the nature of the load profile that can have a higher variation in operational states from one appliance to another. This implies a possible significant difference in shape of a training data set and the test data set of a new appliances not seen by the model in the training set. All other appliances had a high accuracy. The precision of the hot water boiler is lower than the coffee machine. This could be due to a higher variability in the usage duration, which is variable with the amount of water being heated.

TABLE IV
SINGLE APPLIANCE HMM RESULTS ON RAW DATA FOR MULTI-MEDIA APPLIANCES, SAMPLING RATE = 1 SECOND

Appliance category	Appliance type	Accuracy (%)	Precision	f1-score
Multimedia appliances	Screen	99.9	0.99	0.99
	Internet router	100	NA[a]	NA[a]
	Laptop charger	99.6	0.93	0.95
	Television	99.9	0.99	0.99

[a]Not Applicable, Internet router is an always on appliance.

Similar high accuracy results are found for the multi-media appliances regardless of the specificity of one appliance in comparison to another.

Table V shows the accuracy results for all state predictions and for On state predictions for each single appliance HMM when load profiles are pre-processed as described in

TABLE V
SINGLE APPLIANCE HMM RESULTS ON PRE-PROCESSED LOAD PROFILES FOR KITCHEN APPLIANCES, SAMPLING RATE = 1 SECOND

Appliance category	Appliance type	Accuracy (%)	Precision	f1-score
Kitchen appliances	Hot-water boiler	99.9	0.97	0.98
	Refrigerator	99.9	NA[a]	NA[a]
	Coffee-machine	99.2	0.72	0.72
	Washing-machine	99.3	0.99	0.98

[a]Not Applicable, refrigerators are 3-states appliances

II-A3. Using pre-processed load profiles improve the results for washing-machine models significantly. Accuracy is also improved in general for all appliances.

TABLE VI
SINGLE APPLIANCE HMM RESULTS ON PRE-PROCESSED LOAD PROFILES FOR MULTI-MEDIA APPLIANCES, SAMPLING RATE = 1 SECOND

Appliance category	Appliance type	Accuracy (%)	Precision	f1-score
Multimedia appliances	Screens	99.9	0.99	0.99
	Internet router	100	NA[a]	NA[a]
	Laptop charger	94.2	0.99	0.91
	Television	99.9	0.99	0.99

[a]Not Applicable, Internet router is an always on appliance.

Table VI shows the accuracy results for state predictions for each single multimedia appliance HMM on pre-processed load profiles. In comparison to the analysis done on raw data, pre-processing actually decreased the accuracy of detection for the laptop charger.

2) Combined appliances models: Two, three and four appliance models trained on single kitchen appliances are combined. State predictions of the models are evaluated on combined power load profile of two, three and four appliances respectively. Figure 1 shows states prediction of a two-appliances HMM (hot-water boiler and refrigerator).

The accuracy of combined 2, 3 and 4 appliance models can be seen in Figure 2.

C. Data resolution degradation

Optimizing the size of the data storage solution and the transmission volume of data is a major issue when it comes to the cost of implementation in a commercialized product. The minimum resolution required while still guaranteeing a high accuracy is tested by purposefully degrading the data resolution of each profile. Power load profiles used for training and for testing are re-sampled to the following data resolutions: [3 sec, 5 sec, 10 sec, 20 sec, 30 sec, 1 min, 2 min, 5 min, 10 min] to study the impact of the down-sampling on the performances of states prediction.

1) Combined HMM for kitchen appliances: For combined kitchen appliance HMM, On and Off state predictions are presented in Figure 2. Model 2 appliances includes hot water boiler and the refrigerator, model 3 appliances is with the addition of the coffee machine and model 4 appliances includes also the washing machine.

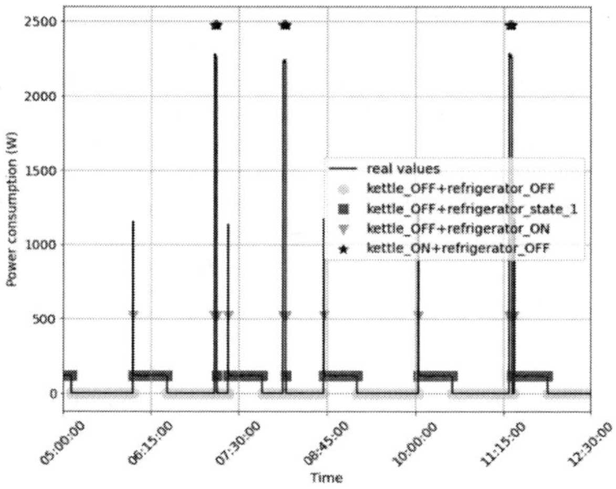

Figure 1. States predictions HMM of hot-water boiler and refrigerator combined, sampling rate = 1 second

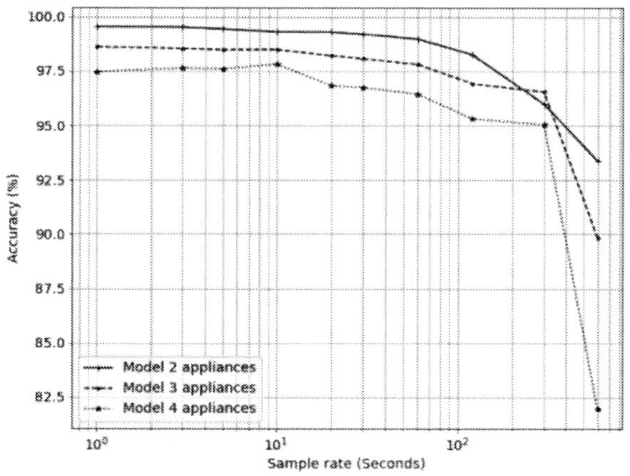

Figure 2. On and Off state prediction accuracy results (%) versus sample rate (Seconds) for combined kitchen appliances models.

For two HMM combined, overall state predictions and On state predictions accuracy is above 98% for every sample rate value bellow 1 minute. Down-sampling data decreases significantly on-states prediction accuracy for combined HMM for a re-sampling rate above one minute :

- for three HMM combined, on-states prediction accuracy decreased from 96.2% to 87.6%,
- for four HMM combined, on-states prediction accuracy decreased from 94.3% to 89.5%.

For combined multi-media device HMM, On and Off state predictions are presented in Figure 3. Model 2 appliances includes the screen and Internet box, model 3 appliances is with the addition of the television and model 4 appliances includes also the laptop charger.

The combined model for multi-media appliances has low accuracy when three and four appliance curves were com-

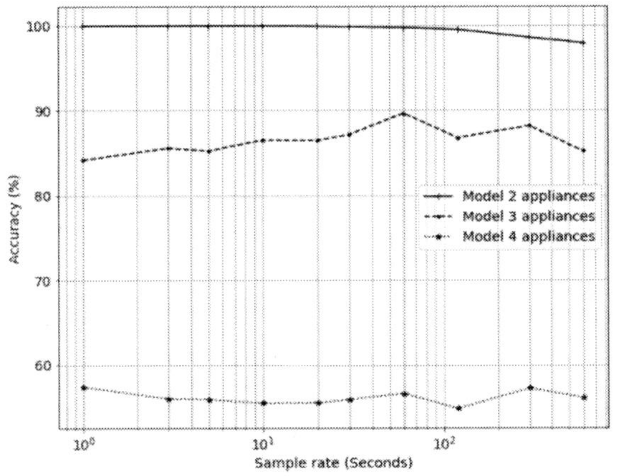

Figure 3. On and Off state prediction accuracy results (%) versus sample rate (Seconds) for combined multi-media device models.

bined. However, the accuracy is relatively constant across all resolutions. This is due to the on off state characteristics of these appliances. The Internet box and laptop charger have a constant power value when plugged in. They also have relatively low power values and the difference between the load curves are very minimal. This is a limitation of this model. The model is incapable of distinguishing and detecting two low magnitude constant power curve signals.

V. Conclusion

This paper has presented a methodology for a HMM applicable to single appliance identification and multi-appliance load curve disaggregation. High accuracy was achieved for all single appliance models. High accuracy was also achieved for multi-appliance models in the kitchen use case. This is due to the unique characteristics of each appliance considered, creating little difficulty for the model to identify all individual states of each appliance in the aggregated load profile. The combined multi-media accuracy was significantly lower than the kitchen appliance use case when 3 and 4 appliance load curves were combined. This was due to the minimal differences between individual load profiles. A constant power profile with low variation and low magnitude also was difficult to desegregate. These characteristics highlight the limits of this model application. Future work could include a hybrid method to apply the HMM only to variable appliance profiles. For example, a linear aggregation model could be used on the constant low consumption appliance profile types which could improve performance for the multi-media disaggregation. The performance of this model could also be compared to existing python packages that have integrated disaggregation algorithms such as the NILMTK python package [17]. The model could be applied to open source data sets as well to compare results with other disaggregation studies such as data sets found in [18].

References

[1] *Statistical office of the european union*, URL: https://ec.europa.eu/eurostat [retrieved: February, 2021], European Commission, 2019.

[2] *Buildings sector energy-related co2 emissions in the sustainable development scenario*, URL: https://www.iea.org/data-and-statistics/charts/buildings-sector-energy-related-co2-emissions-in-the-sustainable-development-scenario-2000-2030 [retrieved: February, 2021], International Energy Agency, 2020.

[3] F. Berg, A. Flyen, Å. L. Godbolt, and T. Broström, "User-driven energy efficiency in historic buildings: A review," *Journal of Cultural Heritage*, vol. 28, pp. 188–195, 2017, ISSN: 1296-2074.

[4] K. C. Armel, A. Gupta, G. Shrimali, and A. Albert, "Is disaggregation the holy grail of energy efficiency? the case of electricity," *Energy Policy*, vol. 52, pp. 213–234, 2013.

[5] T. Alliance, *Projet tbh*, French, [retrieved: February, 2021], 2015. [Online]. Available: https://www.projet-tbh.fr/.

[6] E. CO2, *Picowatty*, [retrieved: February, 2021]. [Online]. Available: https://www.ecoco2.com/plateforme-multi-services-energie-environnement-sante/.

[7] ——, *Awareness raising and accompany residential customers, small companies and local authorities in their energy transition (seize)*, [retrieved: February, 2021]. [Online]. Available: https://www.ecoco2.com/seize-programme-pour-les-economies-denergie-des-entreprises-en-outre-mer/.

[8] C. Klemenjak and P. Goldsborough, "Non-intrusive load monitoring: A review and outlook," Oct. 2016.

[9] A. Faustine, N. H. Mvungi, S. Kaijage, and K. Michael, "A survey on non-intrusive load monitoring methodies and techniques for energy disaggregation problem," *arXiv:1703.00785 [cs]*, Mar. 10, 2017. arXiv: 1703.00785.

[10] J. R. Cuñado and N. B. Linsangan, "A supervised learning approach to appliance classification based on power consumption traces analysis," *IOP Conference Series: Materials Science and Engineering*, vol. 517, p. 012011, Apr. 2019, Publisher: IOP Publishing, ISSN: 1757-899X.

[11] O. Hamid, M. Barbarosou, P. Papageorgas, K. Prekas, and C. Salame, "Automatic recognition of electric loads analyzing the characteristic parameters of the consumed electric power through a non-intrusive monitoring methodology," *Energy Procedia*, International Conference on Technologies and Materials for Renewable Energy, Environment and Sustainability, TMREES17, 21-24 April 2017, Beirut Lebanon, vol. 119, pp. 742–751, Jul. 1, 2017, ISSN: 1876-6102.

[12] V. Abeykoon, N. Kankanamdurage, A. Senevirathna, P. Ranaweera, and R. Udawapola, "Real time identification of electrical devices through power consumption pattern detection," presented at the 2016 First International Conference on Micro and Nano Technologies, Modelling and Simulation, Mar. 1, 2016.

[13] G. W. Hart, "Nonintrusive appliance load monitoring," *Proceedings of the IEEE*, vol. 80, no. 12, pp. 1870–1891, Dec. 1992, Conference Name: Proceedings of the IEEE, ISSN: 1558-2256.

[14] Y. Himeur, A. Alsalemi, F. Bensaali, and A. Amira, "Robust event-based non-intrusive appliance recognition using multi-scale wavelet packet tree and ensemble bagging tree," *Applied Energy*, vol. 267, p. 114877, 2020, ISSN: 0306-2619.

[15] A. Tundis, A. Faizan, and M. Mühlhäuser, "A feature-based model for the identification of electrical devices in smart environments," *Sensors*, vol. 19, no. 11, p. 2611, Jan. 2019, Number: 11 Publisher: Multidisciplinary Digital Publishing Institute.

[16] A. R. et. al., "On the accuracy of appliance identification based on distributed load metering data," presented at the 2nd IFIP Conference on Sustainable Internet and ICT for Sustainability (SustainIT), 2012.

[17] N. B. et al., "Towards reproducible state-of-the-art energy disaggregation," in *Proceedings of the 6th ACM International Conference on Systems for Energy-Efficient Buildings, Cities, and Transportation*, ser. BuildSys '19, New York, NY, USA: Association for Computing Machinery, 2019, pp. 193–202, ISBN: 9781450370059.

[18] S. Kapoor, B. Sturmberg, and M. Shaw, "A review of publicly available energy data sets," The Australian National University, Tech. Rep., 2021.

Grid Benefits from Energy Storage

A Taxonomy for Smart Grid Benefits from Energy Storage

Vivian Sultan
California State University
College of Business and Economics
Los Angeles, CA 90032
Email: vsultan3@calstatela.edu

Abstract—**Grid reliability is one of the greatest challenges facing electric utilities. We argue that energy storage will play a significant role in meeting the challenges facing electric utilities by improving the grid's operating capabilities, lowering cost, ensuring high reliability, and deferring and reducing infrastructure investments. Several studies discuss the benefits of energy storage. This paper offers a taxonomy for smart-grid benefits from energy storage based on previous literature to illustrate four core classes of benefits for the grid. This work provides a solid foundation to equip researchers with the most pertinent information to advance future research in this domain.**

Keywords—peak shaving; load following; intermittent generation; battery; power quality.

I. INTRODUCTION

Thinking of electricity in our homes generally evokes the benefits it offers us. Consumers expect to have uninterrupted access to electricity and are consequently inconvenienced when provisioning is not possible. We like how inexpensive it is. We like that its cost is stable so we do not have to think about what time of day we turn on the dryer or take a shower, even though each of those actions may demand a hefty amount of energy from our power company. Finally, we really like the many life-simplifying gadgets it facilitates, whose numbers seem to increase every year: toaster ovens, hair dryers, computers, mobile computing and communication devices, super-efficient home heating systems, and zero-emission electric cars, to name a few.

In contrast, few of us think much about where the energy comes from or how it gets to our homes. As it turns out, supplying clean, high-quality, inexpensive electricity that is (nearly) always available to meet almost any demand is no mean feat. Utility companies work hard at this, and, although they are getting better at it, they still encounter difficulties. Energy storage, utility-scale batteries, can help considerably.

Considering that energy storage is a critical component for the power network and considering the urgency of energy-storage deployment, this paper addresses the research question: "How will energy storage benefit the electric power network, and what services will it offer grid operators?" The objective of this research is to develop a taxonomy for smart-grid benefits from energy storage, to summarize the core concepts, and to offer a detailed typology for energy-storage types and their characteristics based on the current literature.

The remainder of this article is divided into three sections: Section 2 reviews the literature on utility-scale energy storage. Section 3 offers a taxonomy of energy-storage benefits and discusses those benefits in more detail. Section 4 concludes.

II. LITERATURE REVIEW

We conducted a literature review regarding the smart grid benefits from energy storage to highlight the knowledge base on the topic. One way to grasp the main core of a subject is to look at the references cited in current articles and highlight papers cited repeatedly. This step was particularly helpful in identifying the foundational papers. A literature review is a particularly influential tool in the hands of researchers because it allows scholars to gather and recap all the information about research in a specific field [1].

A. *History of Energy Storage*

Although it is not part of our collective consciousness, energy storage has been used by electric utilities for nearly their entire history. For example, the first major pumped hydroelectric storage plant was built by Connecticut Light and Power in 1929 [2]. The first battery-based utility-scale energy-storage plants were built in the 1980s, including Southern California Edison's Chino Battery Energy Storage Plant offering 10 MW of power and 40 MWh of storage [3].

An August 2013 White House report, written in conjunction with the Office of Electricity Delivery and Energy Reliability, detailed the vital role that energy storage would play in improving grid resiliency and robustness related to weather outages and other potential disruptions [4]. Considering that energy storage is a critical component to be added to the power network and considering the urgency of energy storage deployment, several research studies have discussed the types of energy storage and the smart grid benefits associated with energy-storage technologies.

B. *Types of Energy Storage*

Reference [5] offered an extensive introduction to the wide range of energy-storage technologies, from mechanical systems (flywheels, pumped hydroelectric, compressed air, liquid piston) to superconducting magnets and super capacitors to various chemical energy-storage systems. Pumped hydroelectric energy storage had a significant head start on the field and still retains the edge in both available power (up to 3,000 MW at a single facility) and total energy stored (24 GWh) [6]. However, most of the new research and innovation is in chemical energy storage: hydrogen electrolysis, synthesized methane, and such liquid and dry battery systems as lithium ion, sodium sulfur, and iron chromium [5]. There is good reason for this: While a pumped hydroelectric system can store massive amounts of power and can deliver it at high power levels, like any mechanical system, it suffers from a relatively long response time, on the order of five minutes [7]. While this compares well with other spinning reserves (often idling natural gas or diesel power plants), which can take 10 minutes to produce their rated power [8], most battery system can produce their rated power in less than just a few seconds [9, 10]. In the remainder of this article, *energy storage systems* refers to systems with a response time of seconds or less.

C. *The Smart Grid*

Many of energy storage's benefit stem from enabling the smart grid. Reference [11] offered several definitions of the smart grid, where smart means neat, trim, or intelligent and grid means a network of electrical conductors that distribute electricity to definite points: "An electricity network that can intelligently integrate the actions of all users connected to it, generators, consumers and those that do both, in order to efficiently deliver sustainable, economic and secure electricity supplies" [11, p. 712]. The National Institute of Standards and Technology defined smart grid as "a grid system that integrates many varieties of digital computing and communication technologies and services into the power system infrastructure" [11, p. 712]. The use of distributed energy resources, such as renewable technologies and storages, would facilitate the transition to a smart grid because it can help meet regular power demand [11].

The smart grid would help optimize energy efficiency with a two-way exchange of real-time electricity information between consumers and suppliers. This technology can maximize availability, efficiency, reliability, security, economic and environmental performance, and power distribution. Grid reliability is a very important point that must be studied by researchers since it determines the grid's success in providing the needed services to the users [11, 12].

III. BENEFITS OF ENERGY STORAGE

Several studies discuss the benefits of energy storage. Based on our literature review, We have built a taxonomy for smart grid benefits from energy storage that offers four classes of benefits for the grid: enabling the smart grid, facilitating renewable and intermittent generation, improving transmission and distribution, and increasing grid reliability and power quality. The taxonomy is shown graphically in Figure 1.

Energy storage technologies can enhance the grid's operation and efficiency by quickly responding to changes in demand through the smart grid [12]. Energy storage can be defined as storage with the ability to store a definite amount of energy for the electric grid and to provide the stored energy back to the grid. These storage systems must have a known calendar life in years and cycle life in kWh under identified conditions, maintenance standards with schedules, and round-trip efficiency. Finally, the design for these storage systems could be applied in one or more applications to optimize energy economics and grid operations [12]. According to [13], battery storage is a promising solution to economically improve grid reliability through technology. Other advantages include low self-discharge, high efficiency, and fast response [14].

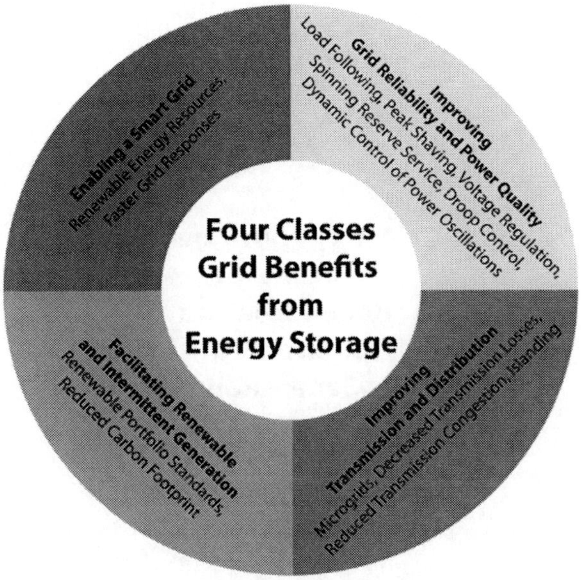

Figure 1. Taxonomy of the grid benefits of energy storage.

A. *Enabling the Smart Grid*

The smart grid is an umbrella term for technologies that are expected to help dramatically increase local and national electricity grid resilience to extreme events, reliability, efficiency, and quality, while reducing their environmental and financial costs. Energy storage can facilitate the smart grid in two ways, by increasing the use of renewable energy resources and by allowing the grid to react more quickly to changes in the operating environment [12]. See Figure 2.

B. *Facilitating Renewable and Intermittent Generation*

Most consumers support moving from carbon-polluting energy sources (such as coal-, diesel-, and natural-gas-fired power plants) to carbon-neutral energy sources (such as solar

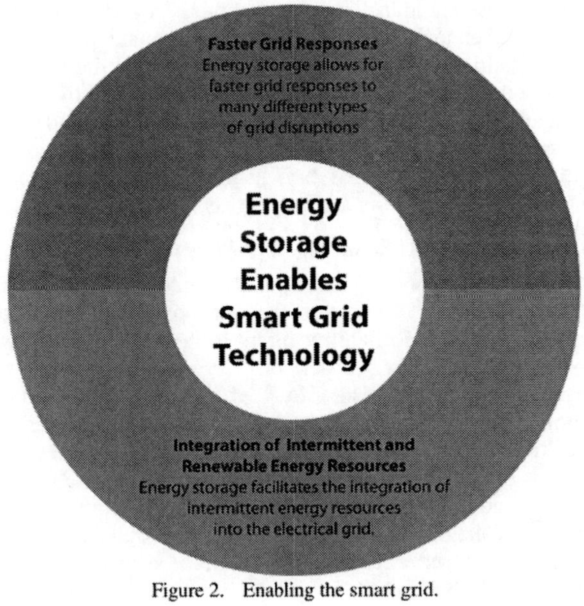

Figure 2. Enabling the smart grid.

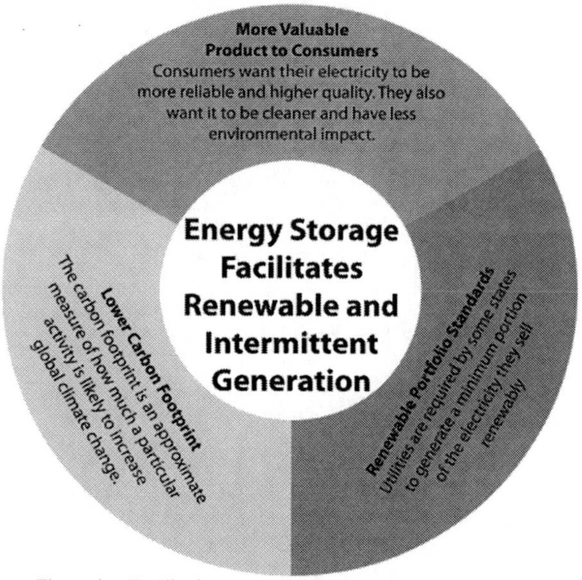

Figure 3. Facilitating renewable and intermittent generation.

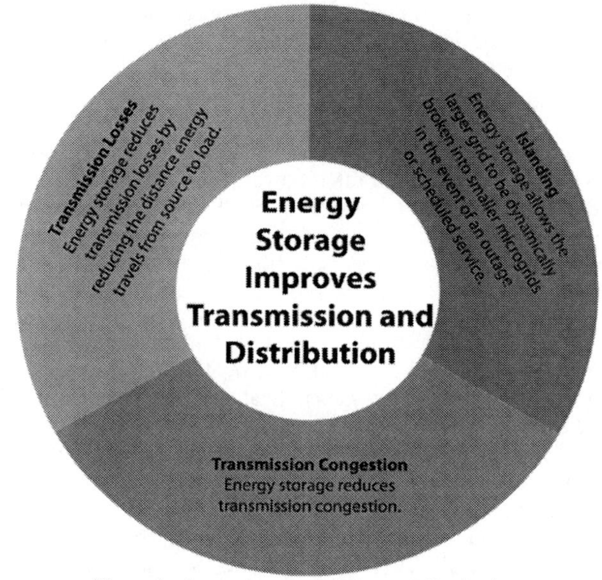

Figure 5. Improving transmission and distribution.

photovoltaics, wind turbines, and run-of-the-river hydro turbines) because it will help reduce climate change and improve local air quality [15]. However, many people believe that these cleaner energy sources' intermittent nature strains the electricity grid's aging technologies ([16–18]. By accepting energy whenever it is available and providing energy when it is needed, energy storage breaks the necessary connection between supply and demand, greatly improving the value of renewable-energy resources to the grid. Properly designed energy-storage systems can help utilities meet states' renewable portfolio standards and reduce their carbon footprints by relying more on carbon-neutral energy sources and less on carbon-polluting sources. See Figure 3.

C. Improving Transmission and Distribution

Even the mundane field of transmission and distribution can be shaken up by energy storage. Long considered a fact of life, transmission losses and congestion can be dramatically improved by the microgrids and islanding enabled by energy storage [19]. Additionally, energy storage system distributed strategically around the grid would allow it to be broken into multiple independent grids during short outages, serving more customers while repairs are in progress [19]. See Figure 4.

D. Improving Grid Reliability and Power Quality

Grid reliability and power quality are the heart of electric utilities' customer satisfaction, the most obvious aspects of the service provided. Reliability refers to how often the power goes out completely. A perfectly reliable gird, where the power never goes out, is a lofty goal, is never attained. In addition to staying on, high-quality power meets customers' expectations in a range of metrics for voltage, frequency, and wave-form regulation. Energy storage helps increase grid reliability and power quality by improving load following, peak shaving, voltage regulation, and droop control, and by offering spinning reserve service and dynamic control of power oscillations [3]. See Figure 5.

Load following means quickly increasing or decreasing the output of an energy source in response to changes in demand. This increase and decrease is important to the grid because it ensures that the right amount of power is available at all times. If too little power is available, the voltage or frequency of the electricity supplied will drop. If too much power is available, at least one of them will increase. In either case, the results can be damage to sensitive electronics or loss of service in some areas. Most energy storage systems are easily ramped up and down in response to the load on the system, improving voltage regulation and droop control [3].

Peak shaving means reducing the highest demand levels at the powerplant. This is important because power produced by peaking plants, those power plants than can quickly ramp up to meet a quick spike in demand, is the most expensive.

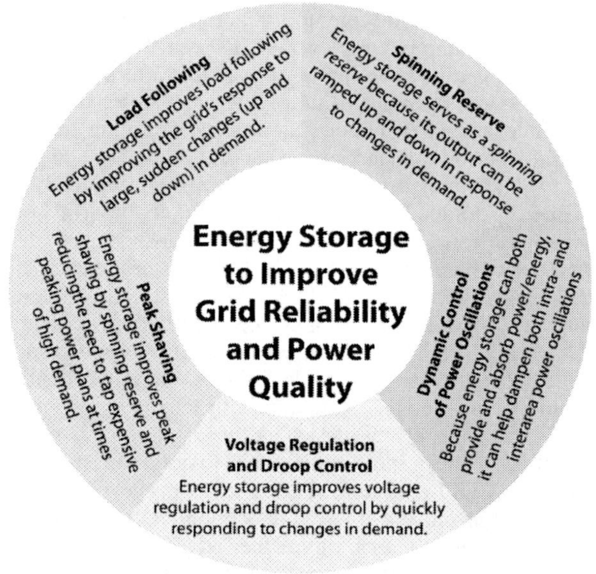

Figure 4. Improving grid reliability and power quality.

Because energy storage systems respond quickly to changes in demand, they provide a spinning reserve service and can dramatically reduce the need to use expensive peaking plants, a savings that utilities can pass on to their customers in the form of lower energy costs [3].

Finally, mismatches between the supply and demand of power can lead to oscillations in the voltage, phase angle, and frequency of the power. These oscillations degrade power quality, possibly damaging sensitive electronic equipment. If that equipment is in the consumer's home or place of business, the consumer is inconvenienced. If that equipment is part of the grid infrastructure, the result can be much worse, possibly leading to a blackout [20].

IV. CONCLUSION

This study aimed to address an important question: "How will energy storage benefit the electric power network and what services will if offer to grid operators?" To answer the research question, we have searched the background literature to develop a taxonomy for smart-grid benefits from energy storage. The taxonomy offered in this paper is the first to address the research question and is intended to help researchers identify areas for future research endeavors.

From this research, we conclude that all four classes of grid benefits are important, but they are too broad to cover in any depth in a single research project. Prospective authors in this space could examine topics such as modeling battery behavior, planning energy-storage backup for the smart grid, battery data management and pipeline to enable analytics, recent advances in battery technology, the future of grid-storage solutions, and advanced designs to integrate energy storage into the electricity systems.

The literature review and taxonomy offered in this paper provide a solid foundation to equip researchers with the most pertinent information. From this taxonomy, a research framework can be developed to provide direction for future research in this domain.

ACKNOWLEDGMENT

The completion of this undertaking could not have been possible without the support of Editide's editing services.

REFERENCES

[1] M. Aneke and M. Wang, "Energy storage technologies and real life applications–A state of the art review," Appl. Energy, vol. 179, pp. 350–377, 2016. Available from: https://doi.org/10.1016/j.apenergy.2016.06.097

[2] F. Cheng et al., "Applying battery energy storage to enhance the benefits of photovoltaics," in 2012 IEEE Energytech. Piscataway, NJ: IEEE, pp. 1–5, May, 2012. Available from: https://doi.org/10.1109/EnergyTech.2012.6304684

[3] H. C. Hesse, M. Schimpe, D. Kucevic, and A. Jossen, "Lithium-ion battery storage for the grid—A review of stationary battery storage system design tailored for applications in modern power grids," Energies, vol. 10, no. 12, art. 2107, 2017. Available from: https://doi.org/10.3390/en10122107

[4] The President's Council of Economic Advisers, the U.S. Department of Energy's Office of Electricity Delivery and Energy Reliability, and the White House Office of Science and Technology (August, 2013). Economic Benefits Of Increasing Electric Grid Resilience To Weather Outages. Available from http://energy.gov/sites/prod/files/2013/08/f2/Grid%20Resiliency%20Report_FINAL.pdf [retrieved: April, 2021]

[5] T. Hou, Y. Cui, X. Zhou, W. Zhang, and B. Ruan, "Charge-discharge strategy for battery energy storage to smooth power fluctuations from wind farms," in 2016 China International Conference on Electricity Distribution. Piscataway, NJ: IEEE, pp. 1–5, August, 2016. Available from: https://doi.org/10.1109/CICED.2016.7576001

[6] S. Keller, "Consumer demand for clean energy significantly increases," Bloomberg Business, April 22, 2019. Available from: https://www.bloomberg.com/press-releases/2019-04-22/consumer-demand-for-clean-energy-significantly-increases [retrieved: April, 2021]

[7] European Energy Research Alliance, "Pumped hydro energy storage," EERA Joint Program SP4—Mechanical Storage, Fact Sheet 1. 2016. Available from: https://eera-es.eu/wp-content/uploads/2016/03/EERA_Factsheet_Pumped-Hydro-Energy-Storage.pdf [retrieved: April, 2021]

[8] C. Campbell, "Making the case for spinning reserve on the grid." Rewewable Energy World. (2011). Available from: https://www.renewableenergyworld.com/storage/making-the-case-for-spinning-reserve-on-the-grid [retrieved: April, 2021]

[9] L. Song, Z. Wang, and W. Luo, "Analysis on the response time of the battery energy storage system." Proceedings of the 2014 International Conference on Materials Science and Energy Engineering, (2015). Available from: https://doi.org/10.1142/9789814678971_0080

[10] X. Fan et al., "Battery technologies for grid-level large-scale electrical energy storage." Trans. Tianjin University vol. 26, pp. 92–103. (2020). Available from: https://doi.org/10.1007/s12209-019-00231-w

[11] S. Lewis, "The 10 largest pumped-storage hydropower plants in the world," Engineering News-Record, April 11, 2018. Available from: https://www.enr.com/articles/44302-the-10-largest-pumped-storage-hydropower-plants-in-the-world [retrieved: April, 2021]

[12] K. Moslehi and R. Kumar, "Smart grid-a reliability perspective," in 2010 Innovative Smart Grid Technologies. Piscataway, NJ: IEEE, pp. 1–8, January, 2010. Available from: https://doi.org/10.1109/ISGT.2010.5434765

[13] Office of Energy Efficiency and Renewable Energy, "History of hydropower," 2016. Available from: https://www.energy.gov/eere/water/history-hydropower [retrieved: April, 2021]

[14] B. P. Roberts, "Deploying battery energy storage in the utility distribution grid," in IEEE PES General Meeting. Piscataway, NJ: IEEE, pp. 1–2, July, 2010. Available from: https://doi.org/10.1109/PES.2010.5589571

[15] G. D. Rodriguez, "A utility perspective of the role of energy storage in the smart grid," in IEEE PES General Meeting. Piscataway, NJ: IEEE, pp. 1–2, July, 2010. Available from: https://doi.org/10.1109/PES.2010.5589870

[16] N. Sharma, and S. Sankar, "Modeling and control of battery energy storage system for providing grid support services," in 2018 Clemson University Power Systems Conference. Piscataway, NJ: IEEE, pp. 1–5, September, 2018. Available from: https://doi.org/10.1109/PSC.2018.8664018

[17] G. Spanos and L. Angelis, "The impact of information security events to the stock market: A systematic literature review," Comput. Secur, vol. 58, pp. 216–229, 2016. Available from: http://doi.org/10.1016/J.COSE.2015.12.006

[18] M. C. Such and C. Hill, "Battery energy storage and wind energy integrated into the smart grid," in 2012 IEEE PES Innovative Smart Grid Technologies. Piscataway, NJ: IEEE, pp. 1–4, January, 2012. Available from: https://doi.org/10.1002/est2.50

[19] M. L. Tuballa and M. L. Abundo, "A review of the development of Smart Grid technologies," Renew. Sust. Energ. Rev., vol. 59, pp. 710–725, 2016. Available from: https://doi.org/10.1016/j.rser.2016.01.011

[20] K. K. Zame, C. A. Brehm, A. T. Nitica, C. L. Richard, and G. D. Schweitzer III, "Smart grid and energy storage: Policy recommendations," Renew. Sust. Energ. Rev., vol. 82, pp. 1646–1654, 2018. Available from: https://doi.org/10.1016/j.rser.2017.07.011

Optimal and Almost Optimal Strategies for Rational Agents in a Smart Grid

Alexander Wallis
Faculty of Interdisciplinary Studies
University of Applied Sciences Landshut
Landshut, Germany
email: alexander.wallis@haw-landshut.de

Sascha Hauke
Faculty of Interdisciplinary Studies
University of Applied Sciences Landshut
Landshut, Germany
email: sascha.hauke@haw-landshut.de

Konstantin Ziegler
Faculty of Interdisciplinary Studies
University of Applied Sciences Landshut
Landshut, Germany
email: konstantin.ziegler@haw-landshut.de

***Abstract*—The rising tide of single household prosumers leads to a paradigm shift for power grid operators. Those prosumers are characterized by their consumption, production and storage capabilities. Via buying and selling electricity, every prosumer becomes a rational agent in the smart grid, trying to maximize one's utility. The optimal short- and long-term behavior can now be analyzed using methods of game theory. In this paper, we present a game theoretic model for smart grids with rational prosumers. Using real-world data, we equipped every agent with a growing class of strategies and then compute the resulting Nash equilibria. The differences in prosumers' utility between optimal and almost optimal strategy selection is given as the *price if not knowing the future*.**

Keywords*—*smart grid; game theory; nash equilibrium; prosumer; battery energy storage system.

NOMENCLATURE

$\mathcal{D}_{\mathrm{C}}^{(a)}$	Daily consumption of agent a.
$\mathcal{D}_{\mathrm{P}}^{(a)}$	Daily production of agent a.
$\ell_{\mathrm{C},t}$	Consumption of agent a at time t.
$\ell_{\mathrm{M},t}$	Market power of agent a at time t.
$\ell_{\mathrm{P},t}$	Production of agent a at time t.
$\ell_{\mathrm{R},t}$	Residual of agent a at time t.
$\mathrm{M}_{\max}$	Maximum price for buying electricity.
M_{HT}	Time interval for high-tariff.
M_{LT}	Time interval for low-tariff.
$m_{\mathrm{buy},t}$	Price for buying at t.
$m_{\mathrm{sell},t}$	Price for selling at t.
$\mathrm{P}_{\max}$	Maximal production power.
$\mathcal{R}_t$	Sum of all agents' residual loads.
$\mathrm{SOC}_{\max}$	Maximal storage capacity.
$\mathrm{S}_{\mathrm{charge}}$	Maximal storage charging power.
$\mathrm{S}_{\mathrm{discharge}}$	Maximal storage discharging power.
$\pi_{\sigma}^{(a)}$	Payoff of agent a for strategy σ.

I. INTRODUCTION

Stable and reliable electricity supply is mandatory for our everyday life. Over the past 100 years, our power grid has been steadily evolving to ensure this supply. Now, with the increasing amounts of electric vehicles and the simultaneous shift of generating process from fossil energy resources to Renewable Energy Resources (RESs), the power grid is facing a Herculean task to maintain this function. The ongoing integration of Distributed Renewable Energy Resourcess (DERs) into the existing power grid, due to its highly volatile behavior, leads to an increase of complexity for network management tasks in terms of stability and security. But not only the power grid is evolving, consumer equipped with production capabilities, e.g., photovoltaic, wind turbine, or diesel generators, are now producer at the same time–so-called *prosumers*.

In [1], we define prosumers, which are capable of producing and storing electricity, as atomic entities in smart holonic micro grids. To achieve stable Smart Grid (SG) operation, especially on low-voltage level, accurate forecasting is necessary to handle the volatile nature of RESs. The increasing capacitiy of battery energy storage systems with the simultaneous decreasing costs, lead to more usage in households may they be residential or commercial. Unfortunately, this introduce more unpredictability into the power grid because every prosumer now has more possible decisions to make. To handle this kind of uncertainty, the study of complex interactions between independent rational actors is needed, which falls in the domain of Game Theory (GT). Therefore, we propose a game of rational prosumers and different kind of electricity markets and give the following contributions:

1) Model for SGs with different kinds of actors represented as rational agents under varying market conditions.
2) Determination of a stable system state by calculating the Nash equilibrium for a finite set of defined strategies.
3) Implementation on a real-world data set and evaluating the game results.

The remaining paper is structured in the following manner. Section II gives an overview of existing research approaches for strategic operation of Battery Energy Storage Systems (BESSs). In Section III, we give a detailed description of the developed model, the agent representation, the different electricity markets, as well as explicit strategy implementations. Afterwards, a game is run on a real-world data set and the system equilibrium is calculated in Section IV. We summarize and discuss the results and conclude with possible improvements and ideas for future works.

II. RELATED WORK

Ensuing from our previous definition of atomic units in a holonic SG, and the need for accurate forecasting models and strategic energy storage operation, we focus on the latter in this paper [1]. Therefore, this section provides an overview of the existing research approaches and their impact on our work. For a clearer description, we break them down into the two main areas of BESS management in general and GT approaches in particular.

A. BESS Management

The efficient integration of BESSs into the existing power grid is a major research topic [2] [3]. The majority of them focus on solving optimization problems like sizing [4] [5] or scheduling [6]–[9]. The authors in [10] propose a optimization model for microgrids with generation capacities, e.g., diesel generator, wind turbine, and one scenario with additional battery storages. Their results show that BESS can assist microgrids in the power generation sector. In [11], an genetic algorithm (NSGA-II) for multi-objective optimization in terms of minimize generation costs and battery life loss is presented. Simulated on two scenarios, abundant and short renewable resources, their method reduces both objectives. One major application of BESSs is peak shaving to reduce peak demand on a power system. In [12] two optimization methods in combination with load forecasting are presented. Furthermore, the authors in [13] also take different electricity tariffs into account to shave and shift peak consumption and conclude that strategic operation can lead to reasonable pay-back investment times.

B. Game Theoretic Approach

One problem of modeling and simulating strategic behavior of an arbitrary number of actors or players within a SG, is the rational thinking and the variable goals of each individual. To overcome this problem, game theoretic approaches gain more attention in recent years [14]–[16]. Basically, GT approaches can be divided into cooperative [17] and non-cooperative games. In [18], the authors propose a cooperative game for sharing storage capacities and the results show effective influence on the power grid. Furthermore, a distributed solution for coalition formation to reduce households' electricity costs within a SG is offered in [19]. Similar, consumers are trading energy with each other to minimize their own electricity bill formulated as a centralized optimization problem in [20]. Another application is the examination of trading mechanism in energy markets. In [21], a detailed review of GT methods for local energy trading scenarios is given.

The previous mentioned related work influence this paper, specifically, led to the consideration of modeling prosumer as rational agents with strategic behavior. Within the scope of this paper, every rational agent tries to maximize their own payoff and never negotiate with other participants in the SG.

III. Model Description

In this paper, we implement prosumer as rational agents within a multi-agent system. To analyze different strategies for buying and selling electricity and also–if available–for charging and dischargin battery storage systems and their respective outcome under varying market conditions, a game theoretic approach is presented. Therefore, we introduce in this section a game (Section III-A) of an arbitrary number of agents (Section III-B) and one specific electricity market (Section III-C). Following this, we define a set of strategies agents can choose from and the selected utility function to evaluate their respective outcome (Section III-D).

A. Game

We propose a game $\mathcal{G}$ to analyze the interactions between a set of players modeled as rational agents and a specific electricity market: $\mathcal{G} = (\mathcal{A}, \mathcal{M}, \mathcal{S})$. Where every agent $a \in \mathcal{A}$ is able to buy and sell electricity from respectively to the market $\mathcal{M}$ based on their chosen strategy $\sigma \in \mathcal{S}$. Figure 1 depicts the information flow between agents and the electricity market. In every time interval $t \in T$, the market sends the price for buying and selling $1\,\mathrm{kW\,h}$ electricity to every agent a within the game. It can be also seen from Figure 1, that no communication between the agents is allowed. This feature is out of scope of this paper but will be implemented in future work to enable coalition formations between agents. Besides buying and selling electricity, a prosumer+ is also able to charge and discharge their storage unit. To handle this additional behavior, a definition and classification of the used agents based on their available actions is given in the following.

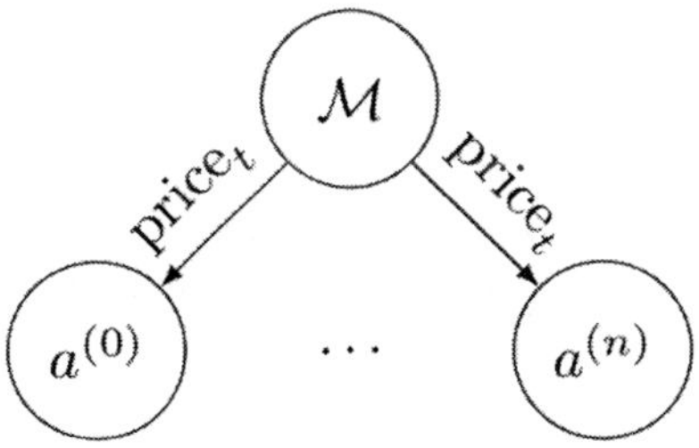

Figure 1. Every agent $a \in \mathcal{A}$ within the game is connected to the same market $\mathcal{M}$ and receive the identical price in every time step t

B. Agent

The goal of this paper is to find the optimal battery charge and discharge strategy based on each individual prosumer's rational utilization. Therefore, every player $a \in \mathcal{A}$ in our game is represented by an rational agent. Agents within our game can be classified into consumer, producer, prosumer with or whitout storage. Depending on this classification, the agents have different properties shown in Figure 2. From Figure 2 can be seen, that a prosumer$^+$agent has a consumption, production, and storage. Furthermore, every agent is connected to the power grid. The daily consumption $\mathcal{D}_{\mathrm{C}}^{(a)}$ of an agent a is fixed in our game and is given by the sum of the consumption in every time interval $\ell_{C,t}^{(a)}$. Similar, the daily production $\mathcal{D}_{\mathrm{P}}^{(a)}$ is the sum of the electricity produced $\ell_{\mathrm{P},t}^{(a)}$ in every time interval

$$\mathcal{D}_{\mathrm{C}}^{(a)} = \sum_{t \in T} \ell_{\mathrm{C},t}^{(a)} \tag{1}$$

$$\mathcal{D}_{\mathrm{P}}^{(a)} = \sum_{t \in T} \ell_{\mathrm{P},t}^{(a)}. \tag{2}$$

For the remaining paper, electricity produced by the production unit or discharged from the storage device is represented by

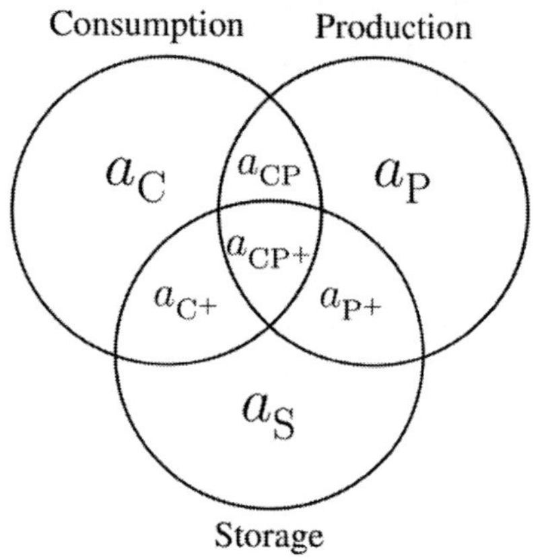

Figure 2. An agent a is composed of at least one of the following properties: Consumption, Production, or Storage

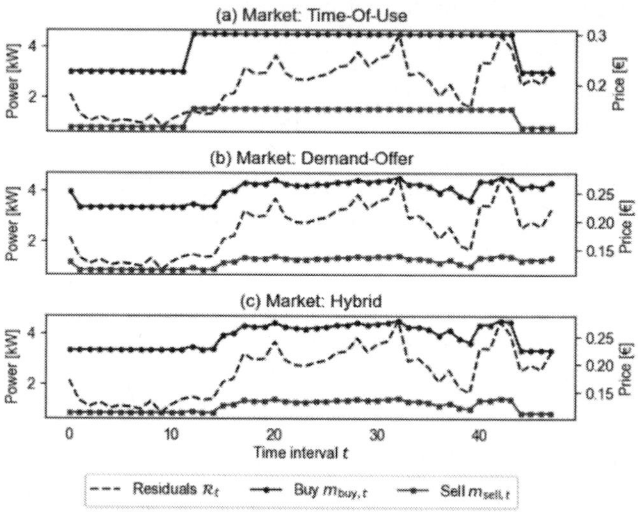

Figure 3. Available different market structures: (a) Time-Of-Use, (b) Demand-Offer, and (c) Hybrid

negative values by convention. Therefore, the provided power in kW per time interval by an agent's production unit with maximum power $\mathrm{P_{max}}$ is given by

$$-\mathrm{P}^{(a)}_{\mathrm{max}} \leq \ell^{(a)}_{\mathrm{P},t} \leq 0. \tag{3}$$

Additionally to the charge and discharge power constraint in (4), the storage unit has also a maximum capacity $\mathrm{SOC_{max}}$ in $\mathrm{kW\,h}$ given in (5). Taken all together, this leads to the state-of-charge calculation of the storage in time interval t in (6).

$$\ell^{(a)}_{\mathrm{discharge},t} \leq \ell^{(a)}_{\mathrm{S},t} \leq \ell^{(a)}_{\mathrm{charge},t} \tag{4}$$

$$0 \leq \mathrm{SOC}^{(a)}_t \leq \mathrm{SOC}^{(a)}_{\mathrm{max}} \tag{5}$$

$$\mathrm{SOC}^{(a)}_t = \mathrm{SOC}^{(a)}_{t-1} + (\ell^{(a)}_{\mathrm{S},t} \times \frac{24\,\mathrm{h}}{T}) \tag{6}$$

The initial daily SOC for every agent is set to the maximum capacity $\mathrm{SOC}^{(a)}_0 = \mathrm{SOC}^{(a)}_{\mathrm{max}}$. We are fully aware of inverter efficiencies at production and storage units, but for convenience and to keep our model simple, we exclude these factors for the scope of the presented work. Since we are focusing on short-term analysis, battery aging due to calendaric or chemical effects are ignored as well. For long-term investing and operating optimization, these factors will be given thorough considerations in future work. Taken all together leads to the main constraint in (7). In every time interval t, an agent's consumption needs to be covered either by produced electricity, discharged storage, or bought from the connected market.

$$\ell^{(a)}_{\mathrm{C},t} + \ell^{(a)}_{\mathrm{P},t} + \ell^{(a)}_{\mathrm{S},t} + \ell^{(a)}_{\mathrm{M},t} = 0 \tag{7}$$

To ensure this constraint, we assume that the power provided by the market/grid is not restricted to any boundaries.

C. Electricity Market

Every prosumer[+] within our game is connected to one shared market $\mathcal{M}$. In this paper, three different types of market structures are used for our model implementation: (I) *Time-of-Use (TOU)*, (II) *Demand-Offer (DO)*, and (III) *Hybrid*. The first market $\mathcal{M}_{\mathrm{TOU}}$ returns the price of buying and selling electricity based on the time of day divided into low-tariff (nighttime) and high-tariff (daytime) givin in (9). In contrast, the second market type $\mathcal{M}_{\mathrm{DO}}$ calculates the price dynamically on the basis of the residual loads (see 8) of every participant in (10). The third market $\mathcal{M}_{\mathrm{Hybrid}}$ is a combination of the other two and calculates the price dynamically within the high-tariff rate.

$$\mathcal{R}_t = \sum_{a}^{\mathcal{A}} \ell^{(a)}_{\mathrm{R},t} \tag{8}$$

$$m_{\mathrm{buy},t} = \begin{cases} \mathrm{M_{max}}, & \text{if } t \in \mathrm{M_{HT}} \\ k \times \mathrm{M_{max}}, & \text{otherwise} \end{cases} \tag{9}$$

$$m_{\mathrm{buy},t} = \mathrm{M_{max}} - \frac{j}{\mathcal{R}_t}, \quad \text{for } \mathcal{R}_t \neq 0 \tag{10}$$

$$m_{\mathrm{buy},t} = \begin{cases} k \times \mathrm{M_{max}}, & \text{if } t \in \mathrm{M_{LT}} \\ \mathrm{M_{max}} - \frac{j}{\mathcal{R}_t}, & \text{otherwise} \end{cases} \tag{11}$$

The price for selling electricity to the market is tied to the one for buying $m_{\mathrm{sell},t} = m_{\mathrm{buy},t} \times l$ with $l \in]0, 1[$, this ensures that rational prosumers prefer self-consumption over selling produced electricity. An example of the daily price calculations for the different markets based on the same residual load is shown in Figure 3.

D. Strategy

In our game, every agent can select a strategy $\sigma \in \mathcal{S}$. We define a strategy σ as a sequence of actions for an agent in every time interval $t \in T$. An action α is a tuple of storage and market operation. Basically, these actions are power values for $\ell^{(a)}_{\mathrm{S},t}$ and $\ell^{(a)}_{\mathrm{M},t}$ after strategy execution. Depending on the agent type and their corresponding properties and actions (see

TABLE I
AGENT CLASSIFICATION AND THEIR ACTIONS

Agent	Property			Action	
	Consumption	Production	Storage	Market	Storage
a_C	√	×	×	√	×
a_{C+}	√	×	√	√	√
a_P	×	√	×	√	×
a_{P+}	×	√	√	√	√
a_S	×	×	√	√	√
a_{CP}	√	√	×	√	×
a_{CP+}	√	√	√	√	√

Table I), unable actions are set to zero. In the following, two concrete strategy definitions are given: (a) SPILLOVER and (b) PRICEDEPENDING(τ)

a) Spillover: This strategy prioritize the storage utilization over selling overproduced electricity. In every time step t, the difference between production and consumption is calculated. This strategy is described in detail in Algorithm 1.

Algorithm 1 Strategy definition for SPILLOVER

Input: Agent a, time step t
Output: Market power $\ell^{(a)}_{\mathrm{M},t}$, Storage power $\ell^{(a)}_{\mathrm{S},t}$
procedure SPILLOVER(Agent a, t)
  $r \leftarrow \ell^{(a)}_{\mathrm{R},t}$
  $s \leftarrow \mathrm{SOC}^{(a)}_{\mathrm{t}}$
  $\ell^{(a)}_{\mathrm{M},t}, \ell^{(a)}_{\mathrm{S},t} \leftarrow 0$
  if $r < 0$ **then**
    $\ell^{(a)}_{\mathrm{S},t} \leftarrow$ CHARGE(s, r) ▷ From Equations (4-6)
    $\ell^{(a)}_{\mathrm{M},t} \leftarrow r - \ell^{(a)}_{\mathrm{S},t}$
  else
    $\ell^{(a)}_{\mathrm{S},t} \leftarrow$ DISCHARGE(s, r) ▷ From Equations (4-6)
    $\ell^{(a)}_{\mathrm{M},t} \leftarrow r - \ell^{(a)}_{\mathrm{S},t}$
  end if
  return $\ell^{(a)}_{\mathrm{S},t}, \ell^{(a)}_{\mathrm{M},t}$
end procedure

It can be seen that the strategy returns values for market and storage power. If an agent doesn't utilize these properties (see Figure I), the strategy return zero for that value.

b) PriceDepending: In contrast to *Spillover*, this strategy focuses the price given by the market. An agent always buys from the market if the price is less than a percentage of the maximum market price $\mathrm{M}_{\max}$ defined by an individual threshold $\tau \in [0, 100]$. Only if this is not the case, an agent charges its storage system at overproduction $\ell^{(a)}_{\mathrm{R},t} < 0$ and discharges it to cover consumption.

To evaluate the different strategies, we define the following utility function

$$\pi^{(a)}_{\sigma} = \sum_t^T (\ell^{(a)}_{\mathrm{M},t} \times c_t)$$
$$c_t = \begin{cases} m_{\mathrm{buy},t}, & \text{if } \ell^{(a)}_{\mathrm{M},t} \leq 0 \\ m_{\mathrm{sell},t}, & \text{otherwise.} \end{cases} \tag{12}$$

Algorithm 2 Strategy definition for PRICEDEPENDING

Input: Agent a, time step t, threshold τ
Output: Market power $\ell^{(a)}_{\mathrm{M},t}$, Storage power $\ell^{(a)}_{\mathrm{S},t}$
procedure PRICEDEPENDING(a, t, τ)
  $r \leftarrow \ell^{(a)}_{\mathrm{R},t}$
  $s \leftarrow \mathrm{SOC}^{(a)}_{\mathrm{t}}$
  $\ell^{(a)}_{\mathrm{M},t}, \ell^{(a)}_{\mathrm{S},t} \leftarrow 0$
  if $m_{\mathrm{buy},t} < \mathrm{M}_{\max} \times \frac{\tau}{100}$ **then**
    $\ell^{(a)}_{\mathrm{M},t} \leftarrow r$
  else
    if $r < 0$ **then**
      $\ell^{(a)}_{\mathrm{S},t} \leftarrow$ CHARGE(s, r) ▷ From Equations (4-6)
      $\ell^{(a)}_{\mathrm{M},t} \leftarrow r - \ell^{(a)}_{\mathrm{S},t}$
    else
      $\ell^{(a)}_{\mathrm{S},t} \leftarrow$ DISCHARGE(s, r) ▷ Equations (4-6)
      $\ell^{(a)}_{\mathrm{M},t} \leftarrow r - \ell^{(a)}_{\mathrm{S},t}$
    end if
  end if
  return $\ell^{(a)}_{\mathrm{S},t}, \ell^{(a)}_{\mathrm{M},t}$
end procedure

An agents' payoff $\pi^{(a)}$ for a specific strategy σ is the sum of money paid or earn in every time step t in (12). Since electricity purchase from the market results in negative costs (see (7)) and vise-versa, the payoff is denoted as the inverse costs. Therefore, a rational agent tries to maximize their resulting payoff. Taken all the previous definitions together, we run in the following a concrete game with specified settings and evaluate the corresponding results.

IV. EVALUATION

After the formal description of the used game model, we evaluate our defined strategies with real-world data described in Section IV-A. Therefore, we present an algorithm for calculating the optimal state for our game–the Nash equilibrium (Section IV-B). After defining a concrete game setup, we present the agents' payoffs under the three previous defined types of electricity markets in Section IV-C.

A. Data Set

For our game implementation, we are using a real-world data set recorded by AUSGRID in the New South Wales (NSW) region in Australia [22]. As part of the *Solar Bonus Scheme* program introduced by the Australian government, electricity consumption and photovoltaic production data from a total of 300 randomly selected residential households were recorded. Altogether, the half hour resolution of the meter data over a time period of three years from 1st July 2010 till 30 June 2013 results in more than 50.000 data points.

B. Nash Equilibrium

To determine the optimal strategy for every agent, we first calculate the Nash equilibrium for the proposed game. In game theoretic approaches the Nash equilibrium is the solution

TABLE II
GAME EVALUATION WITH AGENT SPECIFICATIONS AND RESULTS

Agent	Equipment	Strategy Selection	Market			Price of not knowing the future		
			Demand-Offer	Time-of-Use	Hybrid	Demand-Offer	Time-of-Use	Hybrid
$a^{(0)}$	$C = 290.87\,\text{kW}$	Optimal	-29.45	$-30,49$	$-28,28$	–	–	–
	$P = -73.17\,\text{kW}$	Yesterday	-29.55	$-30,49$	-28.36	0.10	**0**	0.08
	$P_{\max} = 1.7\,\text{kW}p$	Steady	-29.54	-30.49	-28.34	**0.09**	**0**	**0.06**
	$SOC_{\max} = 2\,\text{kW h}$	No Battery	-29.82	-30.86	-28.63	0.37	0.37	0.35
$a^{(1)}$	$C = 151.05\,\text{kW}$	Optimal	-11.19	-11.76	-10.88	–	–	–
	$P = -63.83\,\text{kW}$	Yesterday	-11.43	-11.92	-11.11	0.24	**0.16**	0.23
	$P_{\max} = 1.36\,\text{kW}p$	Steady	-11.20	-11.97	-10.90	**0.01**	0.21	**0.02**
	$SOC_{\max} = 2\,\text{kW h}$	No Battery	-12.73	-13.17	-12.25	1.59	1.41	1.37
$a^{(2)}$	$C = 170.74\,\text{kW}$	Optimal	-14.32	-14.44	-13.57	–	–	–
	$P = -68.73\,\text{kW}$	Yesterday	-14.46	-14.44	-13.70	0.14	**0**	0.23
	$P_{\max} = 1.48\,\text{kW}p$	Steady	-14.37	-14.44	-13.61	**0.05**	**0**	**0.04**
	$SOC_{\max} = 2\,\text{kW h}$	No Battery	-15.56	-15.68	-14.80	1.24	1.24	1.23

where no agent increases their payoff in varying only his strategy unilateral [14]. To reach the equilibrium state in our game, we propose an iterative approach in Algorithm 3 where systematically strategies are ruled out. The set of resulting

Algorithm 3 Iterative Nash calculation

```
Input: Agents 𝒜, Strategies 𝒮, Iterations i
procedure NASH(𝒜, 𝒮, i)
    Initialize Agents A with random Strategy from S
    count ← 0
    while count < i do
        for all a ∈ A do
            P empty list of length |S|
            for all σ ∈ S do
                π(σ) ← CALCULATEPAYOFF(a, σ)
                P ← P + π(σ)        ▷ Append π and σ to list
            end for
            σ_max ← max(P)          ▷ Strategy with max. payoff
            a(σ) ← σ_max            ▷ Set σ_max as agent's strategy
        end for
        count ← count +1
    end while
end procedure
```

strategies for every agent $\sigma^{(a)}$ defines the optimal state within our game. Based on this state, the difference in every agents' payoff for choosing another strategy is calculated and named as the price of not knowing the future.

C. Game Results

For our proposed game $\mathcal{G}$, we choose three different agents $\mathcal{A} = \{a^{(0)}, a^{(1)}, a^{(2)}\}$ from the data set. The set of available strategies is composed of the previously described *Spillover* and *PriceDepending* as well as no battery utilization at all $\mathcal{S} = \{\sigma_{\text{Spillover}}, \sigma_{\text{PriceDepending}(\tau)}, \sigma_{\text{NoBattery}}\}$ with $\tau \in [0, 100]$. Markets are initialized with $k = 0.75, l = 0.5$ and $M_{\max} = 0.30$€. We play the game for a whole week from Monday till Friday. For every day in the week and every market type $\mathcal{M}_{\text{DO}}, \mathcal{M}_{\text{TOU}}, \mathcal{M}_{\text{Hybrid}}$, the optimal strategy based on the previous calculated Nash equilibrium (see Algorithm (3)) is taken as a benchmark. For every agent $a \in \mathcal{A}$, we replay the game with the following three different strategy selection methods (1) *Yesterday*: the agent selects yesterday's optimal strategy; (2) *Steady*: the agent initially selects SPILLOVER and never changes it; (3) *No Battery*: the agent never uses its storage unit at all. Afterwards, we present the difference between the agents' optimal payoffs and the almost optimal payoffs–the so-called *price of not knowing the future*. The specifications of every agents' storage and production capacities as well as the total sum of production and consumption over the whole week are given in Table II. The last three columns correspond to the *price of not knowing the future*, where the lower values are better–indicated in bold font. A final discussion of the results in Table II and an extensive summery follows next.

V. CONCLUSIONS

With the rise of prosumers and the ongoing integration of RESs, the existing power grid is evolving from a centrally managed critical infrastructure to more and more distributed SGs. Prosumers capable of producing electricity are now able to buy from and sell to the market based on their individual rational goals. To study these interactions between actors in a SG and market operators, a agent-based representation of the prosumers with different properties (see Section III-B and Table I) is presented. These properties and their resulting actions are used to define strategies for varying storage unit utilization in Section III-D. Furthermore, different types of electricity markets in terms of price calculation are defined in Section III-C. A dynamic market ($\mathcal{M}_{\text{DO}}$), where every agent influences the price depending on their actual demand. A time based price mechanism, where the tariff is divided into day-tariff and night-tariff ($\mathcal{M}_{\text{TOU}}$) as well as a combination of both types ($\mathcal{M}_{\text{Hybrid}}$). All of this, can be modeled and evaluated in a game between rational prosumer agents and an electricity market.

After the formal definitions, a game composed of three agents, an electricity market $\mathcal{M}$, and a concrete strategy set $\mathcal{S} = \{\sigma_{\text{Spillover}}, \sigma_{\text{PriceDepending}(\tau)}, \sigma_{\text{NoBattery}}\}$ is played.

Therefore, every market type is studied for a stable configuration, the Nash equilibrium, with an iterative Algorithm (3). Divergent from the resulting agents' payoffs, three other strategy selection methods are compared and the divergence is presented–the so-called *price of not knowing the future*. It can be seen from presented results in Table II that no major difference between strategies in the TOU market is noticed with one exception for agent $a^{(1)}$–except at no battery utilization at all. This can be explained by the fact that neither of the proposed strategies exploit this circumstances. For the remaining market types, a *steady* usage of the SPILLOVER strategy is a pretty good choice for an almost optimal strategy–or in reducing the *price of not knowing the future*.

The promising results of our presented game theoretic approach encourage us for further developments and improvments. Possible extensions are implementations of broader strategy spaces or simulating more agents or whole real-world grid structures. Another interesting aspect are long-term analyses in terms of grid stability as well as reliability.

ACKNOWLEDGMENT

The research presented in this paper is supported by the Federal Ministry for Economic Affairs and Energy BMWi (FKZ 03ET1585A) and Bayern Innovativ. The authors would like to thank D. Hehenberger-Risse, J. Straub, N. Körber, and S. Wang for the enlightening discussions.

REFERENCES

[1] A. Wallis, R. Egert, M. Mühlhäuser, and S. Hauke, "A framework for strategy selection of atomic entities in the holonic smart grid," in *10th International Conference on Smart Grids, Green Communications and IT Energy-aware Technologies (ENERGY 2020)*. IARIA, October, pp. 11–16.

[2] M. Faisal, M. A. Hannan, P. J. Ker, A. Hussain, M. B. Mansor, and F. Blaabjerg, "Review of energy storage system technologies in microgrid applications: Issues and challenges," *IEEE Access*, vol. 6, pp. 35 143–35 164, 2018.

[3] J. Barton and D. Infield, "Energy storage and its use with intermittent renewable energy," *IEEE Transactions on Energy Conversion*, vol. 19, no. 2, pp. 441–448, Jun. 2004.

[4] R. Dufo-López and J. L. Bernal-Agustín, "Multi-objective design of PV–wind–diesel–hydrogen–battery systems," *Renewable Energy*, vol. 33, no. 12, pp. 2559–2572, Dec. 2008.

[5] S. X. Chen, H. B. Gooi, and M. Q. Wang, "Sizing of energy storage for microgrids," *IEEE Transactions on Smart Grid*, vol. 3, no. 1, pp. 142–151, Mar. 2012.

[6] H. Morais, P. Kádár, P. Faria, Z. A. Vale, and H. Khodr, "Optimal scheduling of a renewable micro-grid in an isolated load area using mixed-integer linear programming," *Renewable Energy*, vol. 35, no. 1, pp. 151–156, Jan. 2010.

[7] M. Ross, R. Hidalgo, C. Abbey, and G. Joos, "Energy storage system scheduling for an isolated microgrid," *Renewable Power Generation, IET*, vol. 5, pp. 117 – 123, 04 2011.

[8] P. S. Kumar, R. P. S. Chandrasena, V. Ramu, G. N. Srinivas, and K. V. S. M. Babu, "Energy management system for small scale hybrid wind solar battery based microgrid," *IEEE Access*, vol. 8, pp. 8336–8345, 2020.

[9] M. Korpaas, A. T. Holen, and R. Hildrum, "Operation and sizing of energy storage for wind power plants in a market system," *International Journal of Electrical Power & Energy Systems*, vol. 25, no. 8, pp. 599–606, Oct. 2003.

[10] H. Moradi, M. Esfahanian, A. Abtahi, and A. Zilouchian, "Optimization and energy management of a standalone hybrid microgrid in the presence of battery storage system," *Energy*, vol. 147, pp. 226–238, Mar. 2018.

[11] B. Zhao, X. Zhang, J. Chen, C. Wang, and L. Guo, "Operation optimization of standalone microgrids considering lifetime characteristics of battery energy storage system," *IEEE Transactions on Sustainable Energy*, vol. 4, no. 4, pp. 934–943, Oct. 2013.

[12] E. Reihani, S. Sepasi, L. R. Roose, and M. Matsuura, "Energy management at the distribution grid using a battery energy storage system (BESS)," *International Journal of Electrical Power & Energy Systems*, vol. 77, pp. 337–344, May 2016.

[13] G. Barchi, M. Pierro, and D. Moser, "Predictive energy control strategy for peak shaving and shifting using BESS and PV generation applied to the retail sector," *Electronics*, vol. 8, no. 5, p. 526, May 2019.

[14] Z. M. Fadlullah, Y. Nozaki, A. Takeuchi, and N. Kato, "A survey of game theoretic approaches in smart grid," in *2011 International Conference on Wireless Communications and Signal Processing (WCSP)*. IEEE, Nov. 2011, pp. 1–4.

[15] A.-H. Mohsenian-Rad, V. W. S. Wong, J. Jatskevich, R. Schober, and A. Leon-Garcia, "Autonomous demand-side management based on game-theoretic energy consumption scheduling for the future smart grid," *IEEE Transactions on Smart Grid*, vol. 1, no. 3, pp. 320–331, Dec. 2010.

[16] P. Vytelingum, T. D. Voice, S. D. Ramchurn, A. Rogers, and N. R. Jennings, "Agent-based micro-storage management for the smart grid," in *The Ninth International Conference on Autonomous Agents and Multiagent Systems (AAMAS 2010) - Won the Best Paper Award (14/05/10)*, 2010, pp. 39–46, winner of the Best Paper Award Event Dates: May 10-14, 2010.

[17] A. Loni and F.-A. Parand, "A survey of game theory approach in smart grid with emphasis on cooperative games," in *2017 IEEE International Conference on Smart Grid and Smart Cities (ICSGSC)*. IEEE, Jul. 2017, pp. 237–242.

[18] P. Chakraborty, E. Baeyens, K. Poolla, P. P. Khargonekar, and P. Varaiya, "Sharing storage in a smart grid: A coalitional game approach," *IEEE Transactions on Smart Grid*, vol. 10, no. 4, pp. 4379–4390, Jul. 2019.

[19] A. Chis and V. Koivunen, "Coalitional game-based cost optimization of energy portfolio in smart grid communities," *IEEE Transactions on Smart Grid*, vol. 10, no. 2, pp. 1960–1970, Mar. 2019.

[20] N. Yaagoubi and H. T. Mouftah, "Energy trading in the smart grid: A distributed game-theoretic approach," *Canadian Journal of Electrical and Computer Engineering*, vol. 40, no. 2, pp. 57–65, 2017.

[21] M. Pilz and L. Al-Fagih, "Recent advances in local energy trading in the smart grid based on game-theoretic approaches," *IEEE Transactions on Smart Grid*, vol. 10, no. 2, pp. 1363–1371, Mar. 2019.

[22] E. L. Ratnam, S. R. Weller, C. M. Kellett, and A. T. Murray, "Residential load and rooftop PV generation: an australian distribution network dataset," *International Journal of Sustainable Energy*, vol. 36, no. 8, pp. 787–806, Oct. 2015.

Nearshore and Offshore Wind Energy - Potential and Challenges in Vietnam

Nguyen Van Tho
Faculty of Urban-Infrastructure Engineering
Mien Tay Construction University
Vinh Long City, Vinh Long Province, Vietnam
E-mail: nguyenvantho@mtu.edu.vn

***Abstract*—Energy demand for development is increasing worldwide, while fossil energy resources are gradually becoming exhausted, and their use has many negative environmental impacts. The exploitation of potential renewable sources to reduce dependence on fossil fuels is one of the key strategies for national development. Vietnam has a coastline of more than 3000 km, and has a marine area of more than 1 million km^2, with many locations having average wind speeds exceeding 6m/s. The Vietnamese Government's National Power Development Plan for the period from 2011 to 2020 with the outlook to 2030 provides for an increase in installed wind power capacity to around 1 GW by 2020 and to 6.2 GW by 2030. However, insufficient human resources trained in the field of wind power and high investment costs for equipment imported from abroad are among the biggest barriers to wind power development in Vietnam. This paper will discuss the potential and challenges of near and offshore wind energy in Vietnam.**

Keywords-Nearshore and offshore wind energy; wind power; Vietnam.

I. INTRODUCTION

Vietnam has a coastline of more than 3,000 km and a marine area of more than 1 million km^2. The population of Vietnam has increased significantly over the past decades, and now stands at over 96.5 million people [1]. In 2019, Vietnam's economy continued to show fundamental strength and resilience, supported by robust domestic demand and export-oriented manufacturing. Real Gross Domestic Product (GDP) grew by an estimated 7 percent in 2019, similar to 2018, one of the fastest growth rates in the region [2]. Rapid urbanization has also occurred from 2005 in the country [3]. Population growth coupled with increased demand for goods and services, rapid urbanization, and rapidly growing economic activities in the industrial and service sectors are putting pressure on energy sources in Vietnam. According to [2], electricity consumption in Vietnam has tripled over the past decade. Vietnam's power demand is forecasted to increase 8.7%/year on average in the National Power Development Plan (Revised PDP VII), which has replaced the PDP VII in the period of 2016 - 2030. In the coming years, the socio-economic development in Vietnam would require a substantial amount of energy. The projection of future energy demand in Vietnam by [4] is shown in Figure 1.

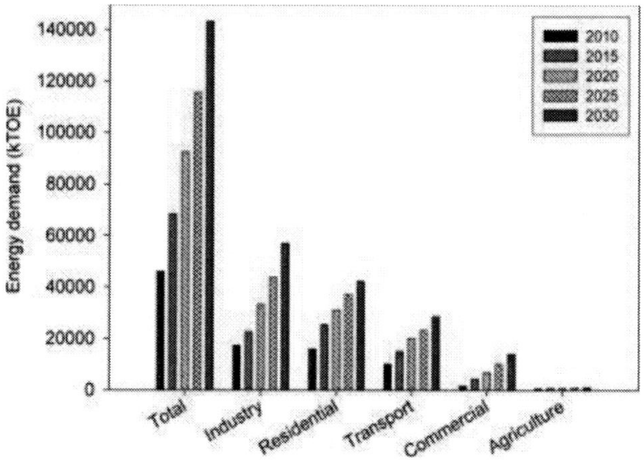

Figure 1. Projection of energy demand in Vietnam [4]

Because of increasing reliance on fossil fuels, the power sector itself accounts for nearly two-thirds of the country's greenhouse gas emissions [2]. Vietnam has a young and fast-growing economy and has put effort into tackling climate change issues at both a national and international level [5]. The Government is promoting the development and use of renewable energy sources, especially wind power, in its strategy to increase domestic energy supply and reduce dependence on fossil fuels. This will contribute to ensuring energy security, climate change mitigation, environmental protection and sustainable socio-economic development. Furthermore, an adequate and reliable energy supply is a necessary prerequisite for sustaining the economic development of the country. Despite the advantage of geographic location and governmental policies in promoting the development of near and offshore wind power, wind power has a long way to go to achieve its potential in Vietnam. The rest of the paper is structured as follows. In Section II, I discuss the potential and challenges of near and offshore wind energy in Vietnam, and offer some conclusions in Section III.

II. POTENTIAL AND CHALLENGES OF NEAR AND OFFSHORE WIND ENERGY

A. *Opportunities and potential*

Vietnam has an advantage of geological location for wind power development. It has a coastline of more than

3,000 km and a marine area of more than 1 million km^2 [1]. The country's sea region has a huge potential for wind energy resources in offshore waters of up to 30m in depth (with an area of 111,000 km^2), and in water of between 30m and 60m in depth (142,000 km^2 in area), which have a potential capacity of producing 64,000 GW and 106,000 GW, respectively. One of results reported in [6] shows that the installed capacity potential is up to 38 GW in only the waters around each Phu Quy or Bach Long Vi island. The coastal area of Binh Thuan - Ca Mau provinces is the most promising region with a potential wind-powered generating capacity of almost 1000 W/m^2 [7]. It was also said that this area has 6 locations with the highest wind speeds (average wind speed of 10 m/s), each with a potential of 1,000 MW and has the advantage of not taking up land, which is a main concern in Vietnam [8]. It is considered one of the regions with high potential globally. Wind farms have been deployed in this region with a total capacity of 1 GW and are expected to be 9 GW in 2030 [7]. In addition, other researchers [9] used a numerical Weather Research and Forecast (WRF) model for studying offshore wind power potential along the southern coast of Vietnam and concluded that the southern coastal area of Vietnam is characterized by strong seasonal and spatial variability in both wind speed and direction. The authors of [9] also indicated that the offshore areas of Ninh Thuan and Binh Thuan provinces are likely to have the largest wind power potential, though still with strong seasonal variation. A recent study [6] reported that wind within 0–185 km from shore throughout Vietnam has the technical potential to generate at least 500 - 600 GW. A wind mapping for Vietnam in 2019 was published by [10]. The technical potential for offshore wind in Vietnam is shown in Figure 2.

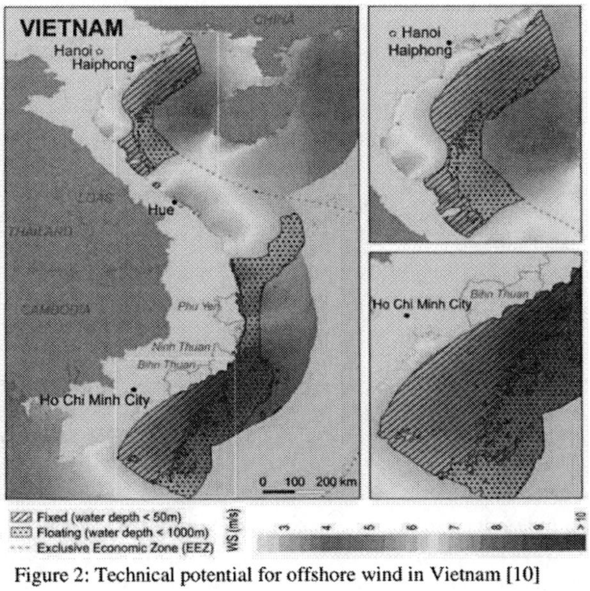

Figure 2: Technical potential for offshore wind in Vietnam [10]

Recognizing this potential, the Government of Vietnam has also issued decisions and policies to promote development in this field, including Decision 1208/2011/QĐ-TTg issued by the Prime Minister on July 21, 2011 approving the National Electricity Development Plan (PDP VII) for the period 2011-2020 with consideration to 2030. In this Decision, the installed wind power capacity needs to increase to around 1 GW by 2020 and to 6.2 GW by 2030 [11]. After the Decision singed in 2011, Decision No.2068/QĐ-TTg was issued by the Prime Minister on November 25, 2015 approving the Renewable Energy Development Strategy (REDS) to diversify the energy supply of Vietnam to the year of 2030, with a vision for 2050 [12]. The REDS emphasizes that the output of electricity produced from wind power will increase from about 180 million kWh in 2015 to about 2.5 billion kWh by 2020; about 16 billion kWh in 2030 and about 53 billion kWh in 2050. It would raise the proportion of total electricity production from wind power from the current negligible level to about 1.0% by year 2020, about 2.7% in 2030 and about 5.0% in 2050. This decision is one of strategies to encourage investment and development of wind power in Vietnam. An example for this is that Vietnam is among leading countries in developing offshore wind power in Asia: China (2.4 GW), Vietnam (0.2 GW), Japan (0.04 GW), South Korea (0.04 GW) and Taiwan (0.008 GW) [13]. In addition, because of Vietnam's wind potential and electricity demand, the Prime Minister issued Decision 39/2018/QĐ-TTg on September 10, 2018 to create a preferential FIT (Feed-in-Tariff) price mechanism to encourage the development of energy types. Currently, the purchase price for onshore wind power according to this Decision is 1,927 VND per kWh, equivalent to 8.5 cent per kWh (excluding VAT), while offshore is 2,223 VND per kWh, equivalent to 9.8 cent per kWh (not including VAT) with projects operating before November 1, 2021 [14]. This Decision has created a new momentum for the wind power market in Vietnam and attracted a series of wind power development projects. As reported by [15], until March 2020, Vietnam had 78 wind power projects with a total capacity of about 4.8GW, which were added to the planning; 11 projects (with total capacity of 377MW) have been operating and generating electricity; 31 projects (total capacity of 1.62GW) have signed power purchase agreements and are expected to be put into operation in 2020-2021. In addition, a further 250 projects with a total capacity of about 45GW have been requested as part of the electricity development plan for Vietnam.

B. *Difficulties and challenges*

There are many difficulties and challenges for near and offshore wind power development. First, lack of human resources trained in the field of wind power is one of the biggest difficulties and challenges in wind power development. On one hand, Vietnam has a wind power project that has been in operation for nearly a decade, and it also has projects that cooperate with international organizations in assessing wind power potential. Consequently, human resources in Vietnam have had the opportunity to access wind energy technology since then (from the wind measurement campaign, detailed design, etc), and currently many wind power projects have been in operation and are under construction. Nevertheless, Vietnam still lacks contractors with experience in installing, repairing and maintaining equipment. Several short-term training courses providing basic knowledge of wind turbine technology and key steps involved in developing wind power projects in Vietnam have been organized mainly at Hanoi University of Science and Technology and Ho Chi Minh City University of Science and Technology. However, there are no specialized majors in the field of renewable energy in general, or wind power in particular, at universities and colleges in Vietnam. Lecturers in this field are lacking and curricula are also limited, creating a shortage of human resources for wind power technology. In addition, shortages of skilled human resources for implementing complete wind power projects including, for example, carrying out assessments of wind resources, preparing investment reports, etc., basic technical and maintenance services, operation and management after installation of wind turbines, are also barriers for development of wind energy in Vietnam [4]. Training and raising professional qualifications for managers, technicians, and skilled workers for wind energy projects should be a priority. An initial step might be to encourage the establishment of specialized energy associations and organizations such as, for example, the Vietnam Wind Power Association and the Renewable Energy Organization, with the participation of experts from prestigious foreign organizations, universities, and research institutes. This mix of expertise could provide advice to the Government and further promote the development of the wind power industry in Vietnam. Second, high investment costs for wind power projects are also among the biggest barriers to wind power development in Vietnam. A successful wind power project often requires a sizable investment in infrastructure and current costs are about 2,500 USD/kW on average, which is equivalent to more than 50 million VND/kW. An example for this is that the Bac Lieu wind power project has had a total investment of about 5,300 billion VND for a total capacity of about 99.2 MW, implemented by Cong Ly Construction - Trade & Tourism Co., Ltd from 2010 to present. This project is supported by an USA bank through a loan guaranteed from the Vietnamese Government. Another difficulty is that wind power technologies are imported from abroad. Wind turbines and other related equipments used for most wind energy projects in Vietnam must be imported from countries such as the USA, Germany, Denmark and so on. In Vietnam, we mainly construct the foundation base and the parts connected to the turbine such as culverts, electrical systems, etc. Therefore, wind power projects depend almost entirely on foreign technology. The Vietnamese government should provide investors with a good opportunity to manufacture wind power equipment so that they can set up factories in Vietnam, or assemble them in Vietnam. This could be done through incentives such as a reduction in tax on assembly of wind power equipment in Vietnam. This would reduce dependence on foreign technology and help reduce the cost of purchasing equipment for wind power generation. Other difficulties and challenges include slow land clearance and a long distance to connect to the power grid [16]. There are also other challenges to wind power development. Environmental problems related to wind power plants such as noise from the turbine blades affecting the daily activities and behavior of animals such as birds, etc. have also been reported [17][18]. Another potential issue is how to dispose of outworn parts of wind turbines safely and without causing environmental pollution. Apart from environmental problems, the connection of wind power plants to any electrical system has impacts on the electricity quality such as variations in voltage, overload in the network, flickers, harmonics, voltage dips, etc. [19]. So, they are also challenges to wind power development in Vietnam in the future. In addition, Vietnam is a developing country and its laws and policies for renewable energy are not stable. This could discourage private and public investment in wind-powered energy projects. For example, [20] reported that from 2007 to 2017, the many Decisions and Circulars for wind energy development promulgated by the Government have hindered the development of wind power projects in Vietnam. More recently, the Decision 39/2018/QĐ-TTg issued on September 10, 2018, as mentioned above, creates a preferential FIT pricing mechanism to encourage wind power development as well as attract more investors, but it also creates difficulties for investors with wind power projects operating after November 1, 2021. According to the proposal of the Ministry of Industry and Trade of Vietnam [15] to the Prime Minister, the wind power purchase and sale price for the projects in operation from November 2021 to December 2022 will be 7.02 cents per kWh for onshore wind power and 8.42 cents per kWh for offshore or nearshore wind power. For projects operating in 2023, the prices are 6.81 and 8.21 cents per kWh, respectively. At present, the impact of the COVID-19 virus is affecting the turbine supply progress, prolonging construction and installation time and slowing down the progress of wind power projects. The production and supply of key equipment, components and accessories of wind power projects are also in shortage and stagnation. The interruptions of the import and export of goods, immigration of technical workers and foreign experts, etc have had negative impacts on enterprises. Investors will face financial difficulties and many influences from Covid-19 if wind power prices fall. They will also encounter the early stage market and the risks of Government's wind power planning adjustment. These could significantly reduce new wind power installations in 2023 and next years.

III. CONCLUSION

The potential for wind power in Vietnam is huge, but it is a relatively new field. Although the Government has issued a range of policies and strategies to develop this energy source, there are still many barriers to the development of this field such as, for example, insufficient human resources, dependence on overseas technology, high investment capital, etc. Developing training programs in this field in universities and research institutes is necessary. Turbines and other related equipment (electrical engine, tower, etc) typically account for a large portion of the total cost of wind power projects (farms). Assembling or manufacturing partly or fully the equipment at domestic factories will help reduce dependence on foreign technology and will reduce costs. Moreover, to implement a successful wind power project, it is necessary to have timely financial support from reputable domestic or international credit institutions and the Government.

ACKNOWLEDGMENT

The author gratefully acknowledges Dr. Barry Clough for very constructive comments and for reading the English manuscript.

REFERENCES

[1] GSO (General Statistics Office of Viet Nam). Statistical Yearbook of Vietnam 2019. Statistical Publishing House, 2020.

[2] WB (World Bank). *Overview*. [Online]. Available from: https://www.worldbank.org/en/country/vietnam/overview, last accessed on 30/04/2021

[3] A. Yamashita. "Rapid Urbanization in Developing Asia and Africa". In: Y. Murayama, C. Kamusoko, A. Yamashita and R. C. Estoque (eds.). Urban Development in Asia and Africa, The Urban Book Series. Springer, Singapore, pp. 47-61, 2017. https://doi.org/10.1007/978-981-10-3241-7_3

[4] N. D. Luong. "A critical review on potential and current status of wind energy in Vietnam". Renewable and Sustainable Energy Reviews 43, 440–448, 2015. http://dx.doi.org/10.1016/j.rser.2014.11.060

[5] D. Nong, C. Wang and A. Q. Al-Aminde. "A critical review of energy resources, policies and scientific studies towards a cleaner and more sustainable economy in Vietnam". Renewable and Sustainable Energy Reviews, 134, 110-117, 2020. https://doi.org/10.1016/j.rser.2020.110117

[6] Vietnam Initiative for Energy Transition. Policy recommendations for offshore wind energy development in Vietnam, 2020. Available from: https://vietse.vn/wp-content/uploads/2020/05/2020-08-28-Offshore-wind-development-Design-EN.pdf, last accessed on 15/05/2021

[7] D. V. Toan, Q. V. Doan, P. L. D. Anh and V. N. Dinh, "The Zoning of Offshore Wind Energy Resources in the Vietnam Sea". In: M. Randolph, D. Doan, A. Tang, M. Bui, V. Dinh. (eds) Proceedings of the 1st Vietnam Symposium on Advances in Offshore Engineering. VSOE 2018. Lecture Notes in Civil Engineering, vol 18. Springer, Singapore, pp. 250–256, 2019. https://doi.org/10.1007/978-981-13-2306-5_34.

[8] EREA & DEA: Vietnam Energy Outlook Report 2019, 2019. https://vietnam.um.dk/~/media/Vietnam/Documents/Content%20English/ENGLISH%20LAYOUT%20FILE%20FULL%202810.pdf?la=en, last accessed on 20/05/2021

[9] V. Q. Doan, H. Kusaka, T. V. Du, D. D Nguyen and T. Con, "Numerical Approach for Studying Offshore Wind Power Potential Along the Southern Coast of Vietnam". In: M. Randolph, D. Doan, A. Tang, M. Bui, V. Dinh. (eds) Proceedings of the 1st Vietnam Symposium on Advances in Offshore Engineering. VSOE 2018. Lecture Notes in Civil Engineering, vol 18. Springer, Singapore, pp. 245–249, 2019. https://doi.org/10.1007/978-981-13-2306-5_33.

[10] ESMAP. Going Global: Expanding Offshore Wind to Emerging Markets. Washington, DC: World Bank, 2019. https://documents1.worldbank.org/curated/en/716891572457609829/pdf/Going-Global-Expanding-Offshore-Wind-To-Emerging-Markets.pdf, last accessed on 15/05/2021

[11] Government of Vietnam. Decision No. 1208/QĐ-TTg dated July 21, 2011 on approving the National Master Plan for Power Development for the 2011–2020 Period with the Vision to 2030 (in Vietnamese). Government of Vietnam, 2011.

[12] Government of Vietnam. Decision No. 2068/QĐ-TTg dated November 25, 2015 approving the Strategy for Renewable Energy Development to Diversify Vietnam's Energy Supply to 2030, with a Vision to 2050 (in Vietnamese). Government of Vietnam, 2015.

[13] H. M. Diaz and G. C Soares, "Review of the current status, technology and future trends of offshore wind farms", Ocean Engineering, Vol. 209, pp. 1-21, 2020. https://doi.org/10.1016/j.oceaneng.2020.107381.

[14] Government of Vietnam, Decision No. 39/2018/QD-TTg dated September 10, 2018 on amending a number of articles of Decision No.37/2011/QD-TTg dated June 29, 2011 of the Prime Minister on the mechanism supporting the development of wind power project in Vietnam (in Vietnamese). Government of Vietnam, 2018.

[15] Ministry of Trade and Industry. Official Document No 2491/BCT-ĐL proposal to prolong the fixed-price mechanism for wind power projects in Decision 39 (in Vietnamese), dated April 9, 2020. Ministry of Trade and Industry, 2020.

[16] H. T. Vo, V. T. Le, L. M. Phung and T. T. H Cao, "Offshore Wind Power in Vietnam: Lessons Learnt from Phu Quy and Bac Lieu Wind Farms". In: M. Randolph, D. Doan, A. Tang, M. Bui, V. Dinh. (eds) Proceedings of the 1st Vietnam Symposium on Advances in Offshore Engineering. VSOE 2018. Lecture Notes in Civil Engineering, vol 18. Springer, Singapore, pp. 276–281, 2019. https://doi.org/10.1007/978-981-13-2306-5_38.

[17] T. A. Kelly, T. E. West and J. K. Davenport, "Challenges and solutions of remote sensing at offshore wind energy developments". Marine Pollution Bulletin 58, 1599–1604, 2009. http://dx.doi.org/10.1016/j.marpolbul.2009.09.002.

[18] S. C. Klain, T. Satterfield, J. Sinner, J. I. Ellis and K. M. A. Chan, "Bird Killer, Industrial Intruder or Clean Energy? Perceiving Risks to Ecosystem Services Due to an Offshore Wind Farm". Ecological Economics 143, 111–129, 2018. http://dx.doi.org/10.1016/j.ecolecon.2017.06.030.

[19] O. Benzohra, S. S. Echcharqaouy, F. Fraija and D. Saifaoui, "Integrating wind energy into the power grid: Impact and solutions". Materials Today: Proceedings 30, 987–992, 2020. https://doi.org/10.1016/j.matpr.2020.04.363.

[20] A. T. Phan, D. Nguyen and T. T. H. Pham, "Research and Development of Wind Power in Vietnam". In: M. Randolph, D. Doan, A. Tang, M. Bui, V. Dinh. (eds) Proceedings of the 1st Vietnam Symposium on Advances in Offshore Engineering. VSOE 2018. Lecture Notes in Civil Engineering, vol 18. Springer, Singapore, pp. 270–275, 2019.https://doi.org/10.1007/978-981-13-2306-5_37.